H. S. YODER, JR.
GEOPHYSICAL LABORATORY
Carnegie Institution of Washington
Washington, D.C.

Generation of Basaltic Magma

NATIONAL ACADEMY OF SCIENCES
Washington, D.C.   1976

Library of Congress Cataloging in Publication Data

Yoder, Hatten Schuyler, Jr., 1921–
  Generation of basaltic magma.

  Based on a lecture series presented at Dartmouth College, May 5-9, 1975, and at the
Geophysical Laboratory of the Carnegie Institution of Washington, June 23-27, 1975.
  1. Basalt—Addresses, essays, lectures. 2. Magmatism—Addresses, essays, lectures.
I. Title.
QE462.B3Y6   552′.2   76-29672

International Standard Book Number 0-309-02504-4

Available from
Printing and Publishing Office
National Academy of Sciences
2101 Constitution Avenue, N.W.
Washington, D.C. 20418

Printed in the United States of America

80 79 78 77 76   10 9 8 7 6 5 4 3 2 1

Preface

A bequest was made by Dr. Arthur L. Day to the National Academy of Sciences "for the purpose of advancing studies of the physics of the earth." The Arthur L. Day Trust Fund Committee* established the Arthur L. Day Prize and Lectureship with the expectation that the recipient would prepare a publication, preferably a book, giving a "comprehensive summary and synthesis of current knowledge in his field."

It was the writer's good fortune to be chosen by the Selection Committee† as the first recipient of the Arthur L. Day Prize and Lectureship, and the opportunity to prepare a book on a subject of his own choosing was indeed a welcome prospect. It had long been his desire to take the time to think through the details of the process of magma‡ generation. Most studies assume the existence of magma and then consider its transport and crystallization; its actual formation is rarely treated at length.

The principal problem was finding sufficiently long periods of time in which to analyze the data and synthesize the views. The bits and pieces

*Trust Fund Committee: William W. Rubey, chairman; Richard M. Goody; Merle A. Tuve; J. Tuzo Wilson.
†Selection Committee: Herbert Friedman, chairman; Jule G. Charney; Allan Cox; Julian R. Goldsmith; O. G. Villard.
‡As used by geologists, magma can be defined as a naturally occurring, mobile liquid within the earth that may contain suspended crystals or rock fragments as well as dissolved or exsolved gases and that can form a rock, crystalline to glassy, on intrusion or extrusion as a lava.

iii

were pulled together over a 2-year period and presented as an integrated lecture series at Dartmouth College on 5–9 May 1975 and at the Geophysical Laboratory of the Carnegie Institution of Washington on 23–27 June 1975. The writer is indebted to Professor Robert W. Decker, Chairman of the Department of Earth Sciences at Dartmouth College, for organizing the visit and to the students and staff for their thought-provoking questions—and answers. The staff of the Geophysical Laboratory provided a stimulating and critical audience for what were intended to be provocative discussions. Their response, substantiated by their willingness to undertake some of the suggested research problems, was most gratifying.

The book is directed to those who would like a comprehensive view of the generation of basaltic* magma at the site of origin.

The writer's purpose was to bring to a focus the many problems amenable to theoretical analysis and experimental investigation. It is evident that direct field study is not possible now, but it cannot be considered as impossible in the near future. The Magma Tap Program, being developed by the Energy Research and Development Administration and the United States Geological Survey in an effort to meet critical energy needs, may present opportunities for the direct observation of at least auxiliary magma chambers. The large uncertainties in the deductions presented here will be evident to the reader, and it would be an understatement to say the points of view are somewhat controversial. If these views stand for any length of time, then the attempt to stimulate new researches on the subject will have been for naught.

The reader will gain an impression of the tenuous nature of the evidence on which major concepts about magmas in the earth are based. The postulated models have become accepted as the reality instead of the lattice of assumptions they are. Authoritarianism dominates the field, and a very critical analysis of each argument is to be encouraged. The structure of facts is minimal—the reader should continuously ask, "What is the evidence?" It is recommended that he withhold judgment on an issue until completing the book. Persuasive arguments in one section may be countered with equal vigor in another

---

*Basalt is a mafic, extrusive igneous rock composed chiefly of plagioclase (usually labradorite) and clinopyroxene. Varieties may contain in addition olivine, orthopyroxene, nepheline, and quartz, individually or in restricted combinations. The rock may be glassy, fine-grained, or porphyritic. Apatite and magnetite are common accessories. Basalt is in some places intrusive in the form of dikes; its intrusive equivalent, however, is generally called diabase (dolerite) or gabbro. The average chemical compositions of the common varieties of basalt have been given by Nockolds (1954), Manson (1976), and Chayes (1975).

section. Not all arguments can be presented, and some are omitted, although referenced, because they are based on experiments since shown to be inaccurate or on assumptions now believed to be unwarranted.

The book is organized about the what, where, how, why, and when of magma production. The lectures included an examination of the critical parameters and principles involved in the generation of the most common magma, basaltic in composition,* that reaches the earth's surface. Constraints are placed on (a) the nature of the parental material and the location of melting in the earth; (b) the melting process, including the heat requirements and sources; (c) the mechanics of magma accumulation, segregation, and rise; (d) the physical chemistry of the evolving magma; (e) the tectonophysics of melting; and (f) the energetics, periodicity, and duration of magmatic events. The substance of the lectures has been expanded somewhat to meet the needs of students having only a general course in petrology.

There is no attempt to discuss subsidiary or auxiliary magma chambers and the associated differentiation processes taking place there, the eruptive mechanics, or the morphological expression of volcanic products. These subjects have been adequately dealt with in texts on volcanology and petrology.

No apologies are made for the breadth of subject matter covered in the book. The writer is obviously not an expert in all the subjects discussed, but who is? An appreciation and awareness of the many exciting facets of magma generation are sought.

It is a pleasure to acknowledge the labors of Miss Marjorie Imlay, who typed drafts and made alterations cheerfully and with patience; Mr. A. David Singer, who organized the drafting, photography, and assembly of the many figures; Mr. William C. Hendrix, who drafted all the figures; and Miss Dolores Thomas, who edited the copy submitted to the National Academy of Sciences and collated the references. An early draft was reviewed by Drs. P. M. Bell (Chapters 1–6) and Bjørn Mysen and Mr. H. Richard Naslund. More advanced drafts were reviewed completely or in part by Dr. Nicholas T. Arndt, Dr. Felix Chayes, Prof. R. W. Decker (Chapters 9 and 10), Prof. John S. Dickey, Dr. Hugh C. Heard (Chapter 9), Dr. Albrecht Hofmann, Prof. Hans G. Huckenholz, Dr. T. Neil Irvine, Dr. Ikuo Kushiro, Prof. John B. Lyons, Prof. Alexander R. McBirney, Prof. S. Anthony Morse, Dr.

---

*Basalts and related rocks (gabbros, amphibolites, and eclogites) comprise about 42.5 percent of the volume of the crust of the earth, according to Ronov and Yaroshevsky (1969, p. 49, Table 7). The base of the crust is taken by them to be the M (Mohorovičić) discontinuity.

Dean C. Presnall (Chapter 6), Dr. Eugene Robertson, Prof. Jean-Guy Schilling, Dr. Jan A. Tullis (Chapter 9), and Dr. Danielle Velde. The many hours invested by these kind friends have led to a much improved manuscript. The debates with these and other specialists and generalists too numerous to list have been most enjoyable and rewarding. As every author knows, the writing of a book on a new subject is a most exhilarating learning experience. All contributors are thanked for their help and tuition.

CONTENTS

vii

1 Introduction

VOLCANIC PRODUCTS—A RENEWABLE RESOURCE

A volcanic eruption is an exciting and terrifying event. It is impossible to transmit in words the total experience of an eruption; each person has different indelible memories of the sights, sounds, vibrations, and emotions. Volcanic eruptions are classed as a geologic hazard. Funding for present-day volcanological studies capitalizes on the emotional fear of the hazard with its resultant loss of life and destruction of property. Only rarely is an eruption recognized for its value—the replenishment of the earth's surface with new material. Farmers appreciate the rich soil formed from some volcanic ashes, and, with suitable crushing equipment, even some new lava flows can become arable within several years. Volcanic products are indeed a principal renewable resource of fundamental importance to mankind.

SYSTEMATIC STUDIES NEEDED—VIOLENT VOLCANOES

Present efforts in volcanological studies are devoted to the short-term events that precede an eruption so that prediction can be made as to the place, time, and intensity of the outbreak. Unfortunately, these studies are made mainly on volcanoes of the quiescent type (e.g., Kilauea, Hawaii), whereas the greatest danger lies with the explosive type (e.g., Lassen Peak, California). Systematic study of the violent-type eruptions should be organized and initiated before another catastrophe in a

1

populated area demands attention. Remote or telemetered sensing devices (e.g., seismic, thermal, tilt, exhalation, photographic) should be installed now at those volcanoes near densely populated areas, particularly in the western United States (e.g., Mount St. Helens; Crandell *et al.*, 1975).

OBSERVED CRYSTALLIZED MAGMA CHAMBERS— TRANSPORTED MAGMAS

As important as the present-day, short-term events are to man, it is necessary to examine the origins of the large volumes of relatively homogeneous lava that issue at the surface. For most purposes it is sufficient to accept the idea that lavas arise from shallow (2–6-km) reservoirs. These reservoirs, after cooling, have been exposed by erosion, and their dimensions can be deduced on the basis of careful field studies—the very substance of geology. Such rock masses, recognized by their characteristics as crystallized from molten or partially molten material, are described as intrusive; that is, the magma was produced somewhere else and was moved, irrespective of the style of emplacement, into the environment in which it is found. Even large batholiths, usually siliceous, multiple, and not visibly floored, are considered intrusive: the magma had moved from its site of origin. Only rarely is the evidence sufficiently clear to lead to the conclusion that small portions of basic or ultrabasic rocks were melted in place (Dickey, 1970; Menzies, 1973). On the other hand, many siliceous rocks exhibit features suggesting *in situ* melting, that is, anatexis, or, more broadly, palingenesis.

MODELS OF MAGMA SOURCES—A PROLOGUE

Because there are few directly applicable data available and the measured parameters commonly lack the sensitivity for the investigator to draw unambiguous conclusions, many different models of magma sources have been proposed. Some of these will be reviewed. There is some risk in commenting on current models of magma generation even though they point up the need for a more thorough examination of the problem. The reader should bear in mind that they are presented out of context and are used only to illustrate some of the principal questions that remain to be answered. It is not appropriate for the writer to criticize them in detail; the reader, however, is not discouraged from that exercise. The models are reviewed for the purpose of arousing the reader's scientific curiosity and to set the stage for what follows.

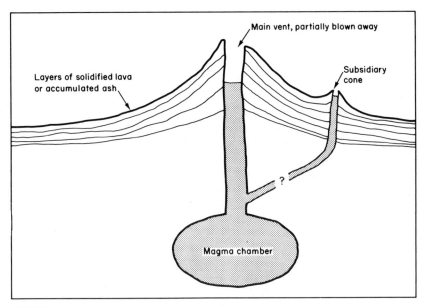

FIGURE 1-1 Cross section of a typical volcano (after Rogers and Adams, 1966, p. 159, Figure 8-3). Shape, size, and position of magma chamber are generally unknown. (With permission of Harper & Row, Publishers.)

BALLOON/SODA-STRAW MODEL

One of the simplest models describing the source region of magma is shown in Figure 1-1. The "balloon/soda-straw" model, no doubt presented to nearly every freshman geology class, implies a point heat source and ellipsoidal isotherms.* Although Rogers and Adams (1966, p. 159, Figure 8-3) indicated that the shape, size, and position were generally unknown, the impressions gained by their relative scales are important. For example, if the vent is 1 km in diameter, the magma chamber center is about 11 km below the summit and the volume of the oblate spheroid chamber is about 90 km³, enough for 900 lava flows of 0.1 km³ in volume. The stored thermal energy in the chamber is about 42×10^{26} erg, five times the total energy of the single, most violent volcanic eruption known.† In the absence of a mechanism for refilling,

---

*Menard (1969, p. 138) gave a similar model with a more spherical chamber situated at much greater depths, just above the asthenosphere, and an appropriately long "soda straw."

†Based on $\Delta E = V \cdot \rho\ (C_P\ \Delta T + \Delta H_m)$, where $V = 90 \times 10^{15}$ cm³, $\rho = 2.78$ g/cm³, $\Delta T = 1250°C$, $\Delta H_m = 87.5$ cal/g, $C_p = 0.25$ cal/g°C (1 cal $= 4.186 \times 10^7$ erg).

it is presumed that the magma would differentiate and deplete with time. Therefore, the chamber obviously could not yield large volumes of homogeneous magma over an extended period of geologic time. Is it possible to drain the chamber completely by successive collapses after each withdrawal? Or is it more likely that the liquid is distributed in a plastic sponge of crystals that deforms as the liquid is squeezed out? What are the forces that cause the magma to rise in the pipe? One must also ask, "What is the meaning of the line bounding the magma chamber?" Is it an isotherm, the solidus, a compositional boundary, a structural discontinuity, a solid-phase change, a diffusion limit, or some other limiting parameter? The difference in the level of the magma in the main vent and that in the subsidiary cone implies a different density of magma; perhaps the magma in the subsidiary cone contains more crystals and therefore is denser on the average.

GEOPHYSICAL MODEL

A more sophisticated, scaled model based on geophysical data collected in Hawaii was given by Eaton and Murata (1960) and is shown in Figure 1-2. The depth of generation was assumed to be that of the deepest earthquake swarms recorded under Kilauea. Eaton and Murata believed that "Such activity appears to mark the zone from which magma is collected and fed into the system of conduits. . . ." The arrows indicate the presumed direction of magma flow into a cavity or at least into a zone of higher permeability and porosity if cavities or pores can be sustained at such depths. The authors emphasized the movement of magma as the cause of quakes but did not relate the magma generation to seismic energy release. The source of the magma was not identified, nor was the thermal regime outlined.

CRYSTAL-MUSH MODEL

At least two additional major ideas are illustrated by the model in Figure 1-3 (Wyllie, 1970, p. 6, Figure 1), based on the description of Green and Ringwood (1967, pp. 164–167). First, the arrows indicate the derivation from an unknown source of a crystal–liquid mush presumed to have an average density less than that of the layer in which it was produced. The question arises as to the ratio of crystals to liquid required for the mush to detach itself from the country rock and move upward. What is the velocity of movement, and is the pressure release sufficient to offset the cooling rate? The second important concept is the depth of separation of liquid from the crystal–liquid mush. No

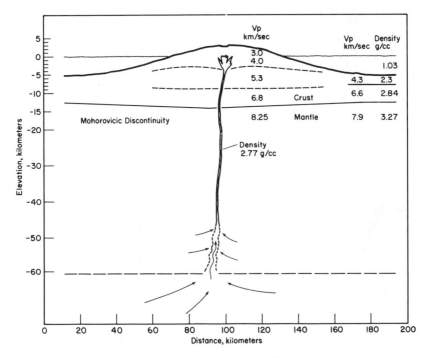

FIGURE 1-2 Schematic cross section of an idealized Hawaiian volcano according to Eaton and Murata (1960, p. 930, Figure 5). (With permission. Copyrighted 1960 by the American Association for the Advancement of Science.)

chemical constraints on where the separation of the liquid takes place are apparent in this model, and Wyllie presumed that the liquid was basaltic irrespective of the depth of separation. Physical constraints, on the other hand, may dominate the process of liquid separation. A zone of shear or a layer of lower density may fix the depth of separation.

BENIOFF-ZONE MODEL

A very imaginative model (Figure 1-4) of the source region of magma was developed by Coats (1962). He had one of the earliest clear pictures of the present-day theory of plate tectonics, emphasizing the Benioff zone (Benioff, 1954) as one of the principal environments of magma generation. He believed that there existed a zone in peridotite, just above 100 km, containing "pockets" of basalt, olivine basalt, eclogite, or preferably vitreous material of unspecified origin that were remelted to yield basaltic liquids or, by the addition of water and materials from the thrust zone, to yield andesite. The quantity of crustal material added to the eruptible basaltic magma determined the extent of "differentiation" along with the basalt–andesite–rhyolite trend. Coats did not specify how the material of basaltic composition initially became molten. What, indeed, were the physicochemical controls that produced the same magma fractionation trend irrespective of the source of materials? His model also illustrates auxiliary

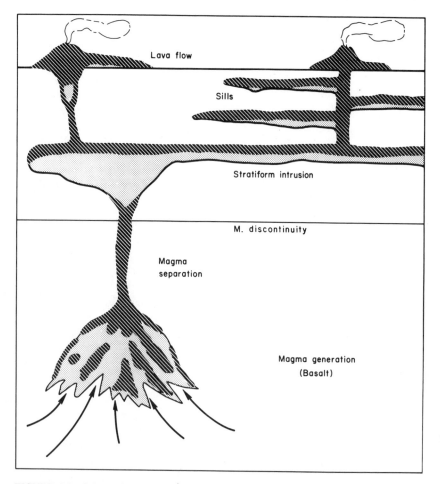

FIGURE 1-3 Schematic representation of magma generation resulting from diapiric uprise of mantle according to Wyllie (1970, p. 6, Figure 1). Stippled areas are crystalline ultramafic material; black areas are interstitial basic liquid or crystalline basalt or gabbro. (With permission of the Mineralogical Society of America.)

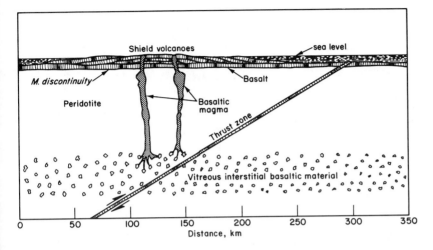

FIGURE 1-4 Diagrammatic cross section through crust and upper mantle of a generalized island arc according to Coats (1962, p. 107, Figure 9), showing only the first stage of development. (With permission. Copyrighted by the American Geophysical Union.)

magma chambers elongated vertically, but the reasons for their existence and elongation were not described. Do the elongated shapes suggest diapirs, or do they result from long-term assimilation of the country rocks? The diagram illustrates basaltic magma originating within a narrow depth range, in contrast to the next model discussed.

DEPTH-OF-GENERATION MODEL

Kuno (1967, p. 108, Figure 10) presented a model (Figure 1-5) that relates the kind of magma (deduced from the belts of rocks observed at the surface) to the locus of earthquakes in the Benioff zone. The earthquakes are present-day events, whereas the volcanoes increase in age from the trough toward the continent (Shimozuru, 1963). The model was constructed in response to a debate with Yoder and Tilley (1962), who believed that Kuno's earlier figure (1959, p. 73, Figure 12) indicated that each magma type was derived from different layers of characteristic parental material. Kuno derived directly, without intervening auxiliary chambers, alkali basalt from the deeper layers, high-alumina basalt from the intermediate depths, and tholeiitic magma from the shallower depths. Attention is called to his interesting characterization of the magma as pockets or droplets, depending on the scale factor, that accumulate into a converging crack system. The localized nature of the melt is also to be noted, even though the thermal regime

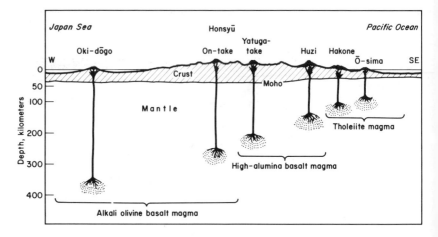

FIGURE 1-5   Cross section of central Honsyū from the Izu Islands to Oki-dōgo Island showing depths of generation of different basaltic magmas according to Kuno (1967, p. 108, Figure 10), who assumed that the mantle was homogeneous and believed that present-day average earthquake foci were applicable. (With permission of Academic Press.)

was not defined. A modification of the model was later suggested by Tarakanov and Leviy (1968), who proposed that each magma originated from special "plastic" layers alternating with "hard" layers. They emphasized the layered structure of the Benioff zone rather than the more widely held view of it as a continuous plate subducted into the mantle.

DEPTH-OF-SEPARATION MODEL

In the same paper, Kuno (1967, p. 108, Figure 11) described the ideas of Yoder and Tilley (1962) in terms of the same environment of magma generation. Although they did not relate magma generation specifically to the Benioff zone, they emphasized the importance of the depth of separation of magma, presumably in auxiliary chambers, from the parental material, as illustrated in Figure 1-6. The magma consisting of crystals and liquid was believed by Yoder and Tilley to originate at relatively uniform depths. A zone having low seismic velocities had been described by Gutenberg (1926), and among the possible interpretations it was considered to be the result of partial melting. Because the liquid in the crystal–liquid mush changes composition as it rises to the surface in response to the change in physicochemical relations, different magma types result, depending on the depth at which the liquid is separated from the mush.

Neither the Kuno model nor his version of the Yoder and Tilley model provides a reasonable explanation for the appearance of alkali basalt as the last event in the formation of some volcanic islands (e.g., Hawaii). It does not seem appropriate for the last magma to come from the greatest depth, usually after a considerable lapse of time, and be required to penetrate the entire volcanic edifice. Note should be taken of those islands (e.g., Lanzarote, Canary Islands; Jan Mayen; Mull, Scotland) where alkali basalts form the shield and the last eruptives are tholeiitic. Although alkaline basalts occurring well behind an island arc do not appear to be restricted in rising from a deep source, those alkaline rocks recently discovered in the zone closest to the associated trench (Arculus and Curran, 1972) may encounter some difficulty in rising to the surface if indeed they are derived from a region directly under the volcano.

EXPLOSIVE-DIATREME MODEL

A composite model was proposed by McGetchin (1975, p. 40) to describe the structure of explosive diatremes and other volcanic fea-

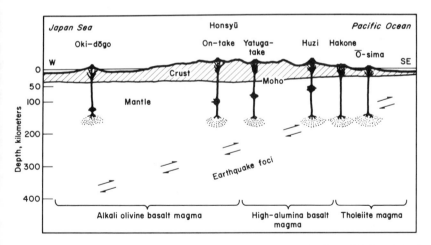

FIGURE 1-6 Cross section of central Honsyū from the Izu Islands to Oki-dōgo Island showing depths of generation of primary magma (stippled area) and sites of liquid separation (irregular black mass on vertical feeder) after Yoder and Tilley's hypothesis as interpreted by Kuno (1967, p. 108, Fig. 11). Not related to present-day earthquake foci in Benioff zone by Yoder and Tilley (1962). (With permission of Academic Press.)

tures (Figure 1-7). He emphasized the importance of gases diffusing (*A*) from depth, thereby fluxing an irregular region (*B*) in a heterogeneous layered mantle. McGetchin shared the view of Jackson and Wright (1970, p. 425, Figure 6), who showed refractory layers intercalated with more fusible layers in a schematic section for Hawaii. The boundary of the region is presumably related to the solidus of the volatile-fluxed rock [McGetchin has agreed with this interpretation (personal communication, 1975)]. It is interesting to speculate whether the crack (*C*) penetrated the molten region or whether it was fortuitously intercepted

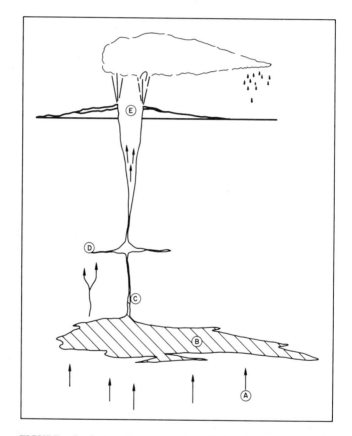

FIGURE 1-7   Composite cross section of volcanic systems according to McGetchin (1975, p. 40, Figure IV-7), showing diffusion of volatiles (*A*), site of magma generation (*B*), dike tapping magma chamber (*C*), auxiliary reservoir (*D*), and explosion crater (*E*). (With permission of Los Alamos Scientific Laboratory.)

by an apophysis of the magma chamber. McGetchin has shown an auxiliary magma chamber (*D*), but it pertains only to quiescent eruptions and is not part of the explosive diatreme system. Compare the size of the surface expression of the diatreme (*E*) and the melt region, recalling that much material in a diatreme is composed of debris from the layers penetrated.

CONCLUSION

The purpose in presenting these models is to illustrate the range of ideas and to give an overview of some of the issues to be examined. Other models will be introduced as appropriate.

2 Parental Materials

PRINCIPAL REQUIREMENT—YIELD BASALTIC MAGMA

To begin the analysis of the problem of magma generation, it is necessary to identify the parental material to be melted because seismic evidence, reviewed below, indicates that, for the most part, the mantle is crystalline. The principal requirement is that the rock or rocks yield a liquid, without crystals, of basaltic composition or a precursor crystal–liquid mush from which such a liquid may be derived at or near the surface of the earth. Basalt has appeared in great volume throughout geologic time and over most of the earth's surface. It is observed today extruding occasionally in an all-liquid condition, and older basalts are found in the glassy state, indicating that they too were once entirely molten. In a broad sense, older and present-day basalts are very similar in composition. Many other major rock types can be related by crystal fractionation, for example, to basaltic compositions (Bowen, 1928), and therefore basalt is usually considered "parental" to other "daughter" magmas.* Of the major rock types, then, basalt

---

*But not *all* igneous rocks need have a basaltic parentage. Andesites, for example, may not necessarily be derivative from basalt; they could form directly from more primitive material (Yoder, 1969, p. 85).

12

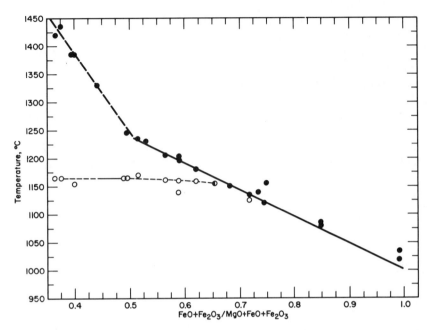

FIGURE 2-1   Plot correlating liquidus temperature with iron enrichment in natural basic rocks determined in the laboratory. Solid dots are liquidus temperatures; open circles mark the temperature where all major phases begin to precipitate together. From Tilley *et al.* (1964, p. 95, Figure 23). (With permission of the Carnegie Institution of Washington.)

should consequently have a higher liquidus* temperature than the daughter products. Because of the great importance of the relationship of liquidus temperature to composition, some details supporting the parent–daughter principle will be given.

PARENT–DAUGHTER TEST

In Figure 2-1 is a plot correlating experimental determinations of liquidus temperatures with iron enrichment in natural rocks. The liquidus temperatures of rocks ranging in iron enrichment from 0.5 to nearly

---

*The liquidus is the maximum temperature of saturation of a solid phase, the primary phase, in a liquid phase for a given bulk composition. Above that temperature the system is completely liquid. In other words, it is the temperature at which the first crystal begins to precipitate from a liquid under equilibrium conditions. In a system of variable composition, the liquidus is the locus of temperature-composition points representing the maximum saturation of a solid phase in the liquid phase.

1.0 can be represented approximately by a straight line. The approximation of a straight line is supported by the determined phase relations in the diopside–forsterite–albite–anorthite system (Yoder and Tilley, 1962, p. 395), where the principal variation is in the plagioclase solid solution. A basaltic magma having an iron enrichment of 0.5 could presumably yield by fractionation all the liquids with successively higher iron-enrichment values. That is to say, a magma with a higher liquidus temperature is a possible parent to daughter magma having a lower liquidus temperature. Conversely, it would be physicochemically unlikely, within reasonable bounds of P_{O_2}, for a magma having a relatively high iron-enrichment value to yield a liquid with a lower iron-enrichment value. The reader may wish to verify this principle by considering the forsterite–fayalite system (Figure 2-2). Similar argu-

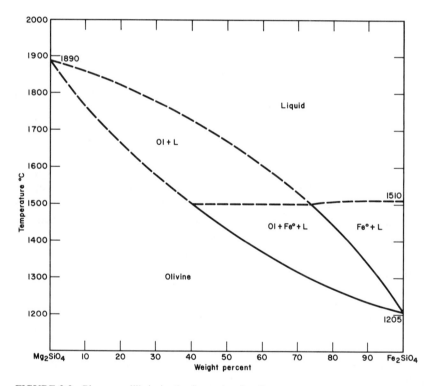

FIGURE 2-2   Phase equilibria in the forsterite–fayalite system at 1 atm [after Bowen and Schairer (1935, p. 163, Figure 7)] with an estimate of the stability of iron (Fe°), a product of the incongruent melting of the iron-rich olivines. (With permission of the *American Journal of Science*.)

ments can be presented for liquidus temperature variations with some other major components (e.g., Al_2O_3/CaO or MgO).

ASSUMPTIONS FOR TEST

Most magmas having an iron enrichment less than 0.5 have accumulated a crystalline phase, commonly olivine, and, therefore, experimentally determined liquidus temperatures are unduly high and not applicable. The group of magmas represented by rocks such as picrites, ankaramites, and oceanites are not found, after natural or dip quenching, wholly in the glassy state. Obviously, one cannot apply the parent–daughter test to rocks formed by the accumulation of crystals. Yoder and Tilley (1962) assumed that the rocks investigated were at one time all liquid and that volatiles were not an important constituent. (As will be seen below, volatiles would lower the liquidus temperature and displace the liquid to a lower derivative status. Rocks having lost their volatiles would melt at anomalously high temperatures in the laboratory.)

The influence of the accumulation of crystals in a magma on the subsequent liquidus temperature measured in the laboratory can be readily appreciated by examining a simple ternary system. In Figure 2-3 the composition P initially crystallizes olivine, which is joined by anorthite when the composition of the liquid reaches L with decreasing temperature. If olivine accumulates in the liquid L, the bulk composition of the accumulated layers might range from P through A to forsterite, depending upon the proportion of trapped liquid. The liquidus temperatures of the crystallized cumulates when measured in the laboratory would be anomalously high relative to that of the parental material P. Remelting such cumulates (Wager *et al.*, 1960) would, of course, produce the same series of derivative magmas as the parental material. Rare olivine-rich liquids such as A (representing picrites or komatiites,* for example) cannot always be dismissed as remelted cumulates, but may be the subsequent products of liquids generated directly by more advanced partial melting of parental materials if the temperature or other conditions are appropriate. In order to obtain the

---

*Thin peridotitic lava flows occur in the Archean Komati formation of Barberton Mountain Land, Republic of South Africa (Viljoen and Viljoen, 1969a,b,c). Quench textures (e.g., spinifex texture) indicate that these ultramafic rocks were essentially all liquid, and the quench textures have been reproduced experimentally from natural samples (Green *et al.*, 1975). The normative olivine content ranges from 55 to 75 wt%. Similar rocks have been found in Ontario and Quebec, Canada (Pyke *et al.*, 1973), and western Australia (Nesbitt, 1971).

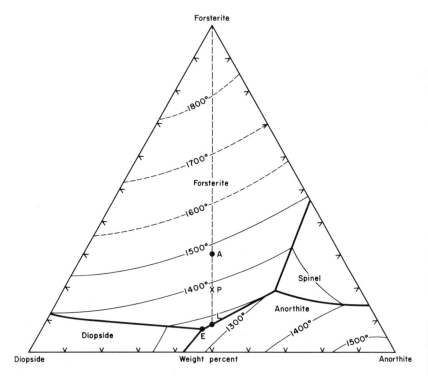

FIGURE 2-3   Phase equilibria in the forsterite–diopside–anorthite system [after Osborn and Tait (1952, p. 419, Figure 5)]. Note change of liquidus temperature as liquid L, derived from bulk composition P, accumulates forsterite. If 23 percent forsterite accumulates in the liquid, the composition of liquid plus accumulated crystals becomes A, and the liquidus temperature is raised 100°C above that of the parental material. (With permission of the *American Journal of Science*.)

necessary high liquidus temperatures, greater than usual depths of generation are envisaged where the mineralogy of the parental material may involve special high-pressure phases (see Figure 4-2) that yield on melting a composition more closely related to such ultrabasic rocks.

BASALTS EQUILIBRATED TO SURFACE CONDITIONS

Yoder and Tilley (1962, pp. 382–383), after examining a large variety of natural basalts at 1 atm, noted that all major phases (olivine, clinopyroxene, orthopyroxene, and plagioclase) can appear as primary silicate phases on the liquidus. They also found that the temperature at which the primary silicate appears lies within a narrow range irrespective of the kind of primary silicate phase. In addition, all major silicate phases appear, with rare exceptions, within a small temperature interval, about 80°C; that is to say, these phases begin crystallizing together at about the same temperature. Yoder and Tilley interpreted these observations to mean that most basalts are eutecticlike* and therefore are products of fractional crystallization. It appears that basalts in a broad sense have equilibrated to the conditions at the earth's surface. In other words, after separation the composition of the liquid has been adjusted successively—presumably by crystallization and removal of appropriate phases—on its rise to the surface so that it is approximately in chemical equilibrium at about 1 atm.

Only rare lavas are believed to have compositions inherited from depth in the mantle. Those alkali basalt lavas containing nodules that have phases (e.g., orthopyroxene) incompatible with the magma at low pressures may exhibit such inheritance. In addition, Kushiro and Thompson (1972, p. 405, Figure 23; see also Kushiro, 1973a, p. 215, Figure 1) showed experimentally that one abyssal olivine tholeiite exhibited eutecticlike behavior at 7.5 kbar and in one interpretation presumed that the liquid had been separated from its parent at a depth of about 25 km. They assumed that no crystallization of the liquid took place en route to the surface and that the composition of the liquid is that of the eutectic of plagioclase peridotite at the determined pressure. These observations are considered to be clear evidence of the physicochemical controls on basaltic magmas. Basalts are not just random products, nor are they themselves primary liquids.

To the best of the writer's knowledge, glass with an iron enrichment of less than 0.5 is exceptionally rare. A few occurrences of such rocks (e.g., komatiites) having unusual quench textures are believed by some to indicate that the rocks were once all liquid. One "fresh" sample having 6.63 H_2O^+ has an iron enrichment value of 0.263 (Green *et al.*, 1975). One possible origin for such rare liquids involves the presence of volatiles, a compositional effect presented in Chapter 4, pp. 78–86.

---

*The eutectic is the lowest melting point of any ratio of components in a system. The proportion of crystalline phases to liquid at the eutectic can be changed by the addition or subtraction of heat without change in temperature. The eutectic is the intersection of the saturation surfaces of the participating crystalline phases in equilibrium with liquid at the lowest melting point. A liquid of the eutectic composition will yield on crystallization under equilibrium conditions all participating crystalline phases at the eutectic temperature if adequate heat is removed. A mixture of participating phases of the eutectic composition will yield on melting under equilibrium conditions all liquid if adequate heat is supplied at the eutectic temperature.

GENERAL TYPES OF SOURCE MATERIALS

The rock types considered as potential source materials for basaltic magma are (a) those that are essentially of basaltic composition and (b) those that are alleged to yield a basaltic magma on partial melting. The

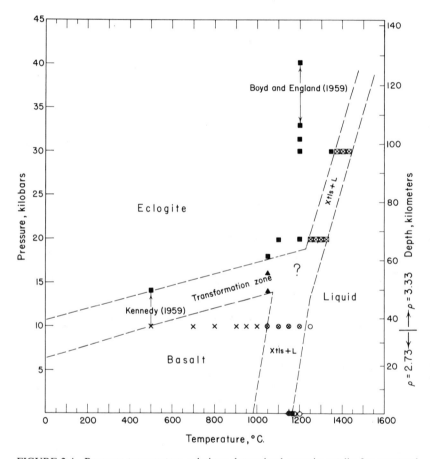

FIGURE 2-4 Pressure-temperature relations determined experimentally for a natural eclogite by Yoder and Tilley (1962, p. 498, Figure 43), a natural basalt converted to glass by Kennedy (1959), and a tachylite by Boyd and England (1959b, p. 88). The symbols mark the conditions of the experiments that define each region. The question mark denotes a region of complex transformations not elucidated by the experiments. Note the restriction of basalts to depths less than 50 km. (With permission of the Oxford University Press.)

first group includes basalt, gabbro, tachylite, amphibolite, hornblendite, and eclogite. Recognition of the importance of olivine in the parental material leads to a consideration of the second group, which includes meteorites, plagioclase peridotites, spinel peridotites, and garnet peridotites. Each of these rock types will be examined in turn.

BASALTIC OR GABBROIC SOURCE

The simplest source of basaltic liquid is a *crystalline* parent of basalt or gabbro. Obviously, it would be necessary to melt the entire parental material to recover the bulk composition desired. Because of crystal settling, fractionation, and the effect of nonhydrostatic stresses, it is not likely that complete melting of a material would ever be achieved *in situ*. If that basalt or gabbro composition is fortuitously at a depth where it corresponds to an invariant (e.g., a eutectic) point, or nearly so, then partial melting will yield the same liquid composition as complete melting (see, for example, phase relations of abyssal olivine tholeiite at 7.5 kbar described by Kushiro, 1973a, p. 215, Figure 1). Such a parental material was suggested as early as 1858 by von Cotta and later by W. L. Green (1887).

Instead of considering a wholly crystalline basalt or gabbro, some petrologists suggested starting with a *liquid* of basaltic composition, a residuum of the processes that generated the earth itself. Daly (1925a) proposed the existence of a *glassy* layer (tachylite), a possibility that has the advantage of requiring essentially no enthalpy of melting. He supposed that the vitreous basalt occurred at a depth of 60 km (≈ 17 kbar), where the temperature was believed to be above the melting temperature of crystalline basalt, perhaps 1300°C. He argued that pressure increased the viscosity of the liquid so that it was a rigid material and responded seismically as a glass. At the temperatures believed to exist in the mantle, however, tachylite ($V_p = 6.5$ km/sec) does not have the appropriate seismic velocities to be considered as a parental material (Adams and Gibson, 1926; Birch and Bancroft, 1942).* Furthermore, it is unlikely that a glass would remain vitreous at the range of temperatures and pressures within the upper

---

*Bridgman (see Daly, 1933, p. 190) obtained a lower compressibility (and therefore a higher velocity is obtained on calculation) on a Hawaiian tachylite, but it contained 20 percent crystals, whereas the tachylite used by Adams and Gibson (1926, p. 276) contained only 3 percent crystals. [Adams and Williamson (1923) gave useful formulae for calculating velocity from the compressibility, β, and the density, ρ, where Poisson's ratio is assumed to be 0.27: $V_p = 13.13(10^6\beta\rho)^{-1/2}$; $V_s = 7.37(10^6\beta\rho)^{-1/2}$.]

mantle. All these suggestions merely delay consideration of the question of the origin of the layers themselves, but there are even more serious objections to having crystalline basalt or gabbro or their glassy equivalents as parental material.

ECLOGITIC SOURCE—HIGH-PRESSURE EQUIVALENT OF BASALT

Yoder and Tilley (1957) found that basalt (and gabbro) are stable only at shallow depths (Figure 2-4), so it is clear that these rocks could be a source for only relatively shallow magmas. It was demonstrated further that all basalts investigated could be converted into eclogite. The eclogites formed experimentally consisted of a simple mineral assemblage, predominantly clinopyroxene and garnet. Some of these eclogites also contained small amounts of hypersthene, quartz, or kyanite, individually or in various combinations. Conversion of the mineral assemblage characteristic of basalt to that of eclogite has been substantiated by experiments on combinations of the simple end-member minerals. In addition to the reactions of anorthite with forsterite and enstatite, discussed below, the breakdown of albite (Birch and LeComte, 1960) and the interaction of clinopyroxene and plagioclase (Kushiro, 1969a, p. 187, Figure 7; Akella and Kennedy, 1971, p. 162, Figure 2) yielded minerals critical to the eclogite assemblage at elevated pressures. The idea that eclogite might be the high-pressure form of gabbro was first put forth by Fermor (1913).

The possibility of eclogite as parental material for basaltic liquids does not present the same difficulty as a pre-existing basalt and gabbro source, where the entire rock must be melted to make basaltic magma. As shown in Chapter 8, eclogites melt over a very narrow temperature range (see Figures 2-4 and 5-5), and each melt fraction yields in general a basaltic composition (Figures 8-5 and 8-6). Whereas there is no objection, disregarding physical problems, to the melting of a rock *in toto*, there would be objection to the melting of that rock if it had a higher melting temperature than its host rock! For example, the model of Coats (Figure 1-4) includes the possibility of pockets of eclogite in peridotite as a magma source, and Ringwood (1969, p. 12, Figure 5) also illustrated magma issuing from eclogite in peridotite below 100 km. As will be seen below, at certain depths eclogite would melt at a higher temperature than its host, for example, garnet peridotite, and therefore would more likely represent a residuum than a source for the liquid.

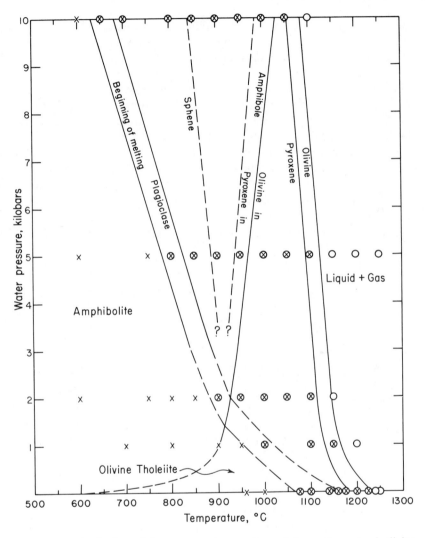

FIGURE 2-5 Projection of the pressure–temperature relations of a natural olivine tholeiite–water system determined by Yoder and Tilley (1962, p. 449, Figure 27). Note the restriction of basalt to low pressures (shallow depths) and the broad field of stability of amphibolite. Symbols mark the conditions of the experiments used to define each significant change in assemblage. × = crystalline; ⊗ = crystals + liquid; ○ = liquid. (With permission of the Oxford University Press.)

AMPHIBOLITIC SOURCE

Because amphibolites have essentially the composition of basalt plus water, they too have been proposed as a source rock for basaltic magmas at high water pressures. Wagner (1928) considered this rock type to be the major parent of basaltic magmas. Yoder and Tilley (1962) demonstrated that amphibolites could indeed be readily made from basalts and gabbros by adding water at high pressures. Furthermore, basaltic liquids in the presence of water can be quenched in the laboratory entirely to amphibole (Figure 2-5), and it is interesting to note that Lacroix (1917) suggested hornblendite as the equivalent of basalt plus water. Unfortunately, amphiboles have their pressure–temperature limits (Gilbert, 1969). Lambert and Wyllie (1970) showed (Figure 2-6) that amphibole was no longer stable at pressures in excess of about 30 kbar (≈100 km).

Not a single residual amphibole has ever been recorded in the entire Hawaiian Island chain even though Tuthill (1969) suggested it as the key mineral in the source rock of those basaltic magmas. Jackson and Wright (1970, p. 414), on the other hand, found amphibole not uncommon in nodules within alkaline basalts, but it was not present in the basalts themselves. A genetic connection has not yet been established between the amphibole-bearing nodules and the basalts. There is no

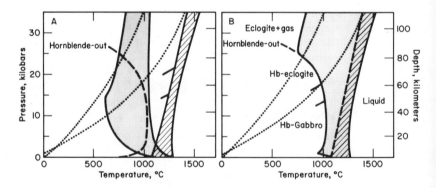

FIGURE 2-6 *A*. Generalized diagram for the melting interval of basaltic material, dry (diagonal ruling) and in the presence of excess water (stippled). The dotted curves are estimated geotherms for the regions under oceans (lower temperatures) and under shield areas (higher temperatures).

B. Estimated melting interval of basaltic material in the presence of 0.1 wt% water. Basic liquid is produced only within the ruled area. Hb = hornblende. From Lambert and Wyllie (1970, p. 765, Figures 1 and 2). (With permission. Copyrighted by the American Association for the Advancement of Science.)

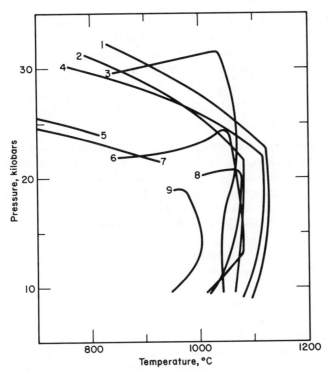

FIGURE 2-7 Experimentally determined stability limits of amphiboles in equilibrium with excess H_2O formed from the following natural samples: (*1*) kaersutite megacryst, Kakanui, New Zealand; (*2*) brown hornblende mylonite, St. Paul's Rocks, equatorial mid-Atlantic ridge; (*3*) olivine nephelinite, Oahu, Hawaii; (*4*) kaersutite eclogite, Kakanui, New Zealand; (*5*) high-alumina olivine tholeiite, Soay, Scotland; (*6*) alkali olivine basalt, Hualalai, Hawaii; (*7*) "alkali olivine basalt," synthetic mixture; (*8*) olivine tholeiite, Kilauea, Hawaii; (*9*) quartz tholeiite, Picture Gorge, Oregon. Summarized by Merrill and Wyllie (1975, p. 564, Figure 7). (With permission of the Geological Society of America.)

evidence of explosive action, such as accompanies volatile release at low pressures, to indicate that amphiboles had broken down on the way to the surface. The breakdown of a wide variety of amphiboles at pressures less than about 30 kbar has since been substantiated by other studies (Figure 2-7). In addition, an amphibolite or hornblendite source of basaltic composition begs the question in the same way as basalt or gabbro does. It is concluded that amphibole does not play a major role in the generation of the most common basaltic magmas, but its presence at depths <100 km may influence the fractionation trends (Bowen, 1928, p. 269–273). Of all the named rock types, only eclogite is still considered a potential source of basaltic magmas, but it lacks what appears to be a key phase, namely olivine.

OLIVINE—AN IMPORTANT SOURCE-ROCK CONSTITUENT?

Glassy basalts with the highest liquidus temperatures at 1 atm have olivine on the liquidus. It appears, therefore, that olivine would probably be a key phase in the parental material. That is, if a derivative magma has olivine on the liquidus, olivine should be an important phase in the parent. This principle does not imply that because olivine is an important phase it is also the most abundant phase. Magnesian olivine has a very high melting temperature; therefore, in systems containing forsterite the olivine usually has a very large liquidus field. This effect usually implies that the composition of liquid at the beginning of melting of the system is displaced by a large amount from forsterite itself. The concept can be illustrated with a simple ternary diagram such as Figure 2-3. If the bulk composition of the parent has olivine on the liquidus (point A or P), one can expect that a liquid of the eutectic composition E removed from the parent would also precipitate olivine on cooling under the same pressure. Alternatives to this concept are presented below. Other arguments in support of the idea that olivine is an important phase in the parental material are given in the following sections.

METEORITIC SOURCE

Silicate-rich meteorites have long been believed to be representative of material comprising the earth's interior. Chondrites, the most abundant (90 percent), are relatively homogeneous in major components. They are probably parts of the solar system and not stray debris from outside it. Their age (4.5 b.y.) is taken to be that of the earth. Some of the isotope ratios, however, are not appropriate, and the abundances of heat-producing elements (K, U, Th) are not wholly consistent with an earth made of chondrite. The K/U is about 8×10^4 for chondrites compared with 10^4 for most terrestrial rocks (Birch, 1965). The $^{87}Sr/^{86}Sr$ is excessively high in chondrites, 0.755, compared with that of terrestrial basalts, 0.705 (Gast, 1960), or oceanic basalts, 0.7026 (Hart, 1971). Some have believed that achondrites are more appropriate representatives of the earth's interior (Washington, 1925, p. 357; Lovering and Morgan,

1964), a view that is supported by recent isotopic data on oxygen (Clayton and Mayeda, 1975). Others have thought that carbonaceous chondrites are more representative because they are the least differentiated meteorites and have more appropriate abundances of the more volatile components (Reed *et al.*, 1960, p. 136).

IMPORTANT PHASES IN METEORITES

All three types of chondrites include members in which olivine is a predominant phase. There are also members rich in orthopyroxene. A petrologist would call such rocks peridotites or, more specifically,

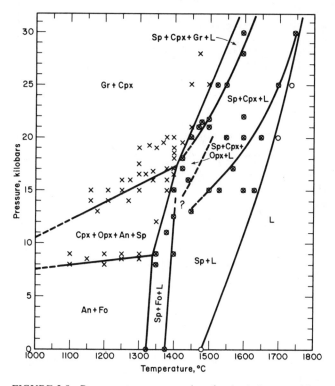

FIGURE 2-8   Pressure–temperature plane for the 1 : 1 composition (molecular ratio) of anorthite and forsterite. Symbols indicate the conditions of the experiments: $\times$ = crystalline; $\otimes$ = crystals + liquid; $\bigcirc$ = liquid. *An* = anorthite; *Cpx* = clinopyroxene; *Opx* = orthopyroxene; *Sp* = spinel; *Gr* = garnet; *Fo* = forsterite; *L* = liquid. From Kushiro and Yoder (1966, p. 340, Figure 1). (With permission of the Oxford University Press.)

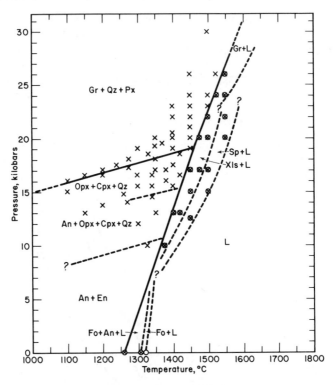

FIGURE 2-9 Pressure–temperature plane for the 1 : 2 composition (molecular ratio) of anorthite and enstatite. Symbols indicate the conditions of the experiments: × = crystalline; ⊗ = crystals + liquid; ○ = liquid. Abbreviations as in Figure 2-8; and *En* = enstatite; *Px* = pyroxene; *Qz* = quartz; *Xls* = crystals. From Kushiro and Yoder (1966, p. 349, Figure 3). (With permission of the Oxford University Press.)

harzburgites. Some members also contain clinopyroxene, but none contains garnet* of a composition appropriate for the formation of basaltic liquids. Most importantly, some chondrites contain plagioclase; therefore, they appear to be low-pressure assemblages. They cannot be products of the *complete* disruption of another large planet; there is, however, the remote possibility that they came from the

*The garnet in the Allende meteorite, for example, is grossularite-rich, whereas the garnets from ultrabasic rocks believed to have been derived from great depths are predominantly pyrope–almandine in composition. Very rare nodules of grospydite from Siberian kimberlite pipes consist of grossularite + clinopyroxene + kyanite (Sobolev *et al.*, 1968).

breakup of the outer shell of a large planet. Presumably, meteorites are condensed matter not incorporated in the aggregation of the primordial planets of the solar system. What then is the evidence that silicate-rich meteorites are low-pressure assemblages? The key relationship is the coexistence of plagioclase with olivine or orthopyroxene.

PHASE INCOMPATIBLE WITH OLIVINE AT HIGH PRESSURE

Anorthite and forsterite react at relatively low pressures (8 kbar), as illustrated in Figure 2-8 (Kushiro and Yoder, 1966, p. 340), to form orthopyroxene + clinopyroxene + spinel. It was also shown that anorthite and enstatite are restricted to relatively low-pressure environments. They react to orthopyroxene + clinopyroxene + quartz at about 15 kbar (Figure 2-9). The presence of albite in a plagioclase would raise the pressure limit of stability (Emslie and Lindsley, 1969, p. 109; K. E. Windom, personal communication, 1975), and iron in an olivine would lower the pressure limit of stability (Green and Hibberson, 1970; Emslie, 1971, p. 155). If meteorites are indeed debris from a disrupted planet, it would have to have been much smaller in diameter than the moon, which has a pressure of 50 kbar at its center. On the

TABLE 2-1 Comparison of Chemical Composition (wt%) of Meteorite (Adjusted) with that of Peridotite

Oxide	Allende, Mexico, Meteorite[a]	Salt Lake Crater, Hawaii, Peridotite[b]
SiO_2	46.50	48.26
TiO_2	0.22	0.22
Al_2O_3	4.56	4.90
Cr_2O_3	0.70	0.25
FeO	9.91	9.94
MnO	0.24	0.14
MgO	33.31	32.51
CaO	3.59	2.98
Na_2O	0.60	0.66
K_2O	0.04	0.07
P_2O_5	0.32	0.07
TOTAL	99.99	100.00
FeO/(FeO + MgO)	0.229	0.234

[a] After subtraction of the following elements by Seitz and Kushiro (1974, p. 954, Table 1): O_2, 4.73 wt%; Fe, 19.2 wt%; Ni, 1.45 wt%; S, 2.21 wt%.
[b] Kuno and Aoki (1970, p. 276, Table 2, No. 24) recalculated after subtraction of H_2O^-.

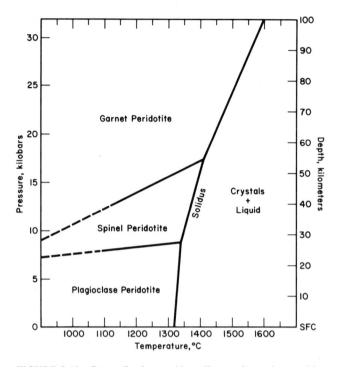

FIGURE 2-10   Generalized assemblage diagram for various peridotites based on experimental data in Figure 2-8. Depth scale based on
average density of 3.2 g/cm³. *SFC* = surface.

basis of Fe/Mg partitioning between olivine and orthopyroxene,
silicate-rich meteorite assemblages, even though brecciated, generally
appear to be in equilibrium.* The olivine of meteorites, ranging in
composition from Fo_{85} to Fo_{70}, would be suitable as a constituent of a
parental material capable of yielding a residue having a modal olivine of
Fo_{90}, but meteorites would not be suitable for yielding liquids with
normative Fo_{90}. [The reader may again wish to refer to the forsterite–
fayalite phase diagram of Bowen and Schairer (1935, p. 163, Figure 7)
revised in Figure 2-2.] The prevailing oxidation state would be ex-

*Van Schmus and Wood (1967, p. 757, Table 2, i) have based a classification of
chondrites on the degree of equilibration of olivine and orthopyroxene; however, the
samples displaying a lack of homogeneity are believed to be few (B. Mason, personal
communication, 1976).

tremely important, and meteorites are known to have a very wide range of oxidation.

METEORITES RELATED TO PERIDOTITE?

The compositions of silicate-rich meteorites can be adjusted on reasonable grounds so that their norms* are comparable to those of peridotite found on earth (Table 2-1). It is necessary to remove metallic Fe, Ni, and S, as well as oxygen (except from some reduced chondrites). (If meteorites are representative of the material condensed to form the earth, then the Fe, Ni, and S are presumably now in the core.) The existence of meteorites rich in olivine provides major support for the argument that peridotite is the parental material in the earth. On the other hand, the presence of plagioclase and the absence of garnet, a major contributor to melts of basaltic composition at high pressure, would appear to constitute serious objections to the acceptance of silicate-rich meteorites *per se* as mantle material unless metamorphosed. On the basis of the experiments summarized in Figures 2-8 and 2-9, it was demonstrated that meteorites are low-pressure assemblages, and it now remains to be shown that the meteorites are related to various other types of peridotite found in the earth. Figure 2-10, based on the work of Kushiro and Yoder (1966) and MacGregor (1968), illustrates schematically the relationship of plagioclase peridotite to spinel peridotite and garnet peridotite.

Spinel peridotites are found as large intrusions in the central parts of fold belts, as nodules in alkali basalts, and as small intrusions in large fault zones. Garnet peridotites are commonly found in fold belts subjected to high-grade metamorphism and as nodules in alkali basalts and kimberlite pipes. Garnet peridotites have a relatively narrow range of composition and are of worldwide occurrence. It would appear from Figure 2-10 that garnet peridotites are stable only at depth in the upper mantle. One can obtain an appreciation of just how deep in the mantle these garnet peridotites are derived by examining the compositions of the coexisting minerals.

---

*The norm is the calculated mineral composition of a rock in terms of hypothetical standard mineral end members similar, especially in basalts, to those of the actual modal minerals that occur or would form on complete crystallization from the same chemical constituents at relatively low pressures in the absence of volatiles. The norm was devised by Cross, Iddings, Pirsson, and Washington in 1902 and is commonly referred to as the CIPW system, after the first letter of the last name of each author. The method of calculation may be found in Washington (1917, pp. 1162–1164), Holmes (1921, pp. 427–431), Johannsen (1939, pp. 88–92), and other petrological works.

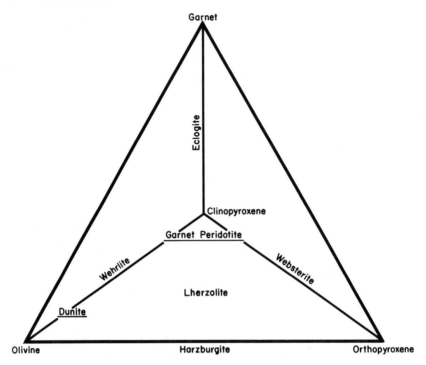

FIGURE 2-11   Nomenclature tetrahedron for assemblages of olivine, clinopyroxene, orthopyroxene, and garnet. Rock names underlined lie within the tetrahedron.

GARNET PERIDOTITES FROM THE UPPER MANTLE

The principal minerals of garnet peridotites are olivine, orthopyroxene, clinopyroxene, and garnet. The compositions of these phases may be displayed in a tetrahedron, and appropriate names may be assigned to their various combinations (Figure 2-11). The compositions of the coexisting pyroxenes change as a function of temperature, and the solvus relating these changes, determined experimentally by Davis and Boyd (1966) and Boyd (1970), is relatively insensitive to pressure (Mysen, 1976a). In addition, the alumina content of the orthopyroxene in the presence of *garnet* is a function of pressure,* and that relationship has been determined experimentally by MacGregor (1974). The relationship of the alumina content of orthopyroxene to pressure is also sensitive to temperature, so that the temperature must be estimated first from the two-pyroxene solvus. These indirect indicators of pressure

---

*The alumina content of orthopyroxene in the presence of *spinel* does not appear to be sensitive to pressure (Obata, 1975, p. 2, Figure 1; Fujii and Takahashi, 1976).

and temperature* are considered to be relatively insensitive to minor elements (e.g., Ti: Akella, 1974).

With this information, Boyd and Nixon (1973) have tested the pyroxenes in nodules of garnet peridotite from kimberlite pipes in which the nodules are thought to have been carried up rapidly from great depths in explosive events. The pressure and temperature deduced from each nodule are assumed to define the conditions under which the nodule last equilibrated. Also in the pipes is a large array of other assemblages related to garnet peridotite: for example, harzburgite, lherzolite, websterite, dunite, and rarely eclogite. Similar nodules are also found in alkali basalts but are exceptionally rare in tholeiites. The results that Boyd and Nixon obtained from nodules in the pipes of northern Lesotho, shown in Figure 2-12, indicate that the nodules record conditions of very high pressures and temperatures of equilibration. Metamorphic and metasomatic effects would tend to decrease the magnitude of these values. Although the interpretation of the abrupt change in the curvature of some of the pyroxene geotherms is being debated (e.g., Mercier and Carter, 1975; Goetze, 1975), the range of physicochemical conditions registered by the nodules from the upper mantle is of great importance.

KEY POSITION OF GARNET PERIDOTITE

It is convenient to illustrate the assemblages found in the nodules schematically in the $CaO–MgO–Al_2O_3–SiO_2$ system (Figure 2-13). The diagram brings to light the key position of garnet peridotite (Fo + En + Di + Py) relative to the eclogite assemblage (Di + Py), which provides most of the components for basaltic liquids, as well as other peripheral assemblages recorded mainly as nodules from kimberlite pipes. Several questions arise: (a) Are these assemblages representative of the primary mantle itself? (b) Are they the cumulates from liquids produced on partial melting of the mantle? Or (c) are they the residua from the melting of the mantle? It does not appear to be possible at present to reach an unambiguous decision on the basis of chemistry alone†; however, other tests are presented below (pp. 107–109).

*Other estimates of the temperature or pressure of formation of garnet peridotite can be made using the Fe–Mg partitioning between the following mineral pairs after further calibration:

    Opx-Cpx (Kretz, 1961)
    Opx-Gr (Wood and Banno, 1973)
    Ol-Cpx (Obata *et al.*, 1974; Powell and Powell, 1974)
    Cpx-Gr (Råheim and Green, 1974; Akella and Boyd, 1974)

†An example of the graphical techniques for relating parental material, primary liquid, residua, and cumulates is given by O'Hara *et al.* (1975, p. 585 ff.).

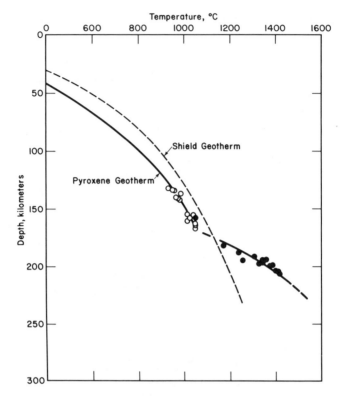

FIGURE 2-12   Estimates of temperatures and depths of equilibration of lherzolite nodules (granular, open circle; sheared, filled circle)
primarily from kimberlites of northern Lesotho. Temperatures were
estimated from the diopside solvus of Davis and Boyd (1966) using the
Ca/(Ca + Mg) ratio of the natural diopsides. Pressures were estimated from the data of MacGregor (1974) using the raw Al_2O_3 content
of the natural enstatites. From Boyd and Nixon (1973, p. 432, Figure
1). (With permission of the Carnegie Institution of Washington.)

WILL GARNET PERIDOTITE YIELD BASALTIC MAGMA?

If the reader accepts the conclusions that (a) garnet peridotite has the
appropriate mineralogy for the conditions of the upper mantle, (b) the
bulk composition is consistent with the source materials forming the
earth, and (c) the rock is prominent among the recovered deep-seated
samples, it is necessary then to ask whether garnet peridotite is
capable of yielding material of basaltic composition.

Unfortunately, only one garnet lherzolite and one spinel lherzolite
have been studied at a series of pressures and temperatures in the

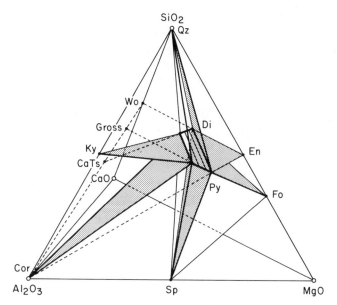

FIGURE 2-13 The CaO–MgO–Al₂O₃–SiO₂ system, illustrating the position of the eclogite assemblage (Py–Di tie lines), representing basaltic compositions, and its relationship to the presumed parental garnet peridotite assemblage (Fo + En + Di + Py). One type of solid solution each is illustrated for Di and Py, but none of the solid solutions for other end-member compounds are shown. Stippled planes are major ternary assemblages separating quaternary assemblages, . which are Di + Py + Ky + Qz; Di + Py + Ky + Cor; Di + Py + Cor + Sp; Di + Py + En + Qz; Di + Py + En + Fo; and Di + Py + Fo + Sp. *Fo* = forsterite; *En* = enstatite; *Di* = diopside; *Py* = pyrope; *Sp* = spinel; *Cor* = corundum; *CaTs* = Ca-Tschermak's molecule; *Ky* = kyanite; *Gross* = grossularite; *Wo* = wollastonite; *Qz* = quartz. End-member compounds plotted in mol percent. After O'Hara and Yoder (1967, p. 69, Figure 1).

laboratory. Ito and Kennedy (1967) found that a garnet lherzolite transformed to an assemblage consisting of olivine + clinopyroxene + orthopyroxene + spinel at pressures less than 20 kbar. A spinel lherzolite assemblage persisted at pressures less than 20 kbar, according to the experiments of Kushiro *et al.* (1968a), and transformed to garnet-bearing assemblages above 20 kbar. The run times may have been too short for equilibrium to be attained or the investigated temperatures too high for the solidus of basalt to be reached, but the main observation is that no plagioclase was seen. In view of the data of Figures 2-8 and 2-9, the lack of plagioclase reaction relations at 1 atm, and the lack of unequivocal data on the natural rocks, the absence of plagioclase suggests that (a) these lherzolites do not have the requisite composition to produce a basalt or (b) the basaltic components had already been removed—that is, the rock was depleted of basaltic components by partial melting, for example, and is therefore a residuum.

Alternatively, the normative An and Ab of the potential plagioclase at 1 atm may be incorporated modally as the calcium Tschermak's molecule and jadeite molecule, respectively, in clinopyroxene. Clinopyroxenes may contain up to 26 wt% normative An at 1 atm, according to Yoder and Tilley (1962, p. 406, Table 19). The normative An content of the garnet lherzolite investigated by Ito and Kennedy (1967, p. 521, Table 1, No. 2) is 4.7 wt%, whereas the spinel lherzolite investigated by Kushiro et al. (1968a, p. 6024, Table 1) had a normative An content of 10.2 wt%. Furthermore, a synthetic peridotite close to Nockolds' (1954) average with 8 percent normative An yielded in the crystal + liquid region a glass of basaltic composition by chemical analysis (Reay and Harris, 1964). Plagioclase was not observed as a phase in those experiments at 1 atm. Assuming that equilibrium was achieved, there appears to be a dilemma: How can one derive basalt, whose principal phase is plagioclase, from a rock composition that does not precipitate plagioclase at low pressures, especially 1 atm?

A sheared garnet lherzolite was partially melted (10–25 percent liquid) by Kushiro (1973b, p. 297, Table 85), and the analyzed glass was basaltic in normative character in spite of the fact that no plagioclase was observed (Table 2-2). It appears that partial melting of some undepleted peridotites can yield a liquid of basaltic composition if the liquid is separated from the crystals. If the liquid is not separated from the crystals, then the peridotite will not necessarily yield plagioclase, which presumably remains occult in clinopyroxene.

METEORITES DO NOT YIELD BASALTIC MAGMA

Some peridotites appear to yield liquids of basaltic composition on partial melting,* yet closely related chondrites, which have aggregated

---

*Peridotites that retain all the constituents required to form *and* yield a liquid of basaltic (or eclogitic) composition are often referred to as *undepleted*. A peridotite from which have been extracted some of or all the constituents required to form a liquid of basaltic composition is considered by some investigators to be *depleted*. The applicability of the terms to a given peridotite depends on the past history of the rock, depth, temperature, and composition(s) of the basaltic liquid to be extracted. The terms are ambiguous and should be explicitly defined if used.

TABLE 2-2   Compositions (wt%) of Glasses Formed by Partial Melting of a Natural Garnet Lherzolite (Kushiro, 1973b)

Oxide	Starting Material[a]	10 kbar, 1375°C	15 kbar, 1450°C	20 kbar, 1475°C	Chilled Bronzite Gabbro[b]	Eutectic Composition at 40 kbar, $Di_{47}Py_{47}Fo_3En_3$
SiO_2	44.54	50.3	51.4	50.3	51.33	50.16
Al_2O_3	2.80	12.4	14.1	12.3	13.69	11.89
TiO_2	0.26	1.07	1.24	1.10	1.01	—
Cr_2O_3	0.29	0.44	0.31	0.38	n.d.	—
FeO^c	10.24	10.5	10.0	10.4	10.17	—
MnO	0.13	0.19	0.21	0.18	0.18	—
MgO	37.94	9.62	8.07	9.97	8.94	25.78
CaO	3.32	12.5	11.2	12.0	11.60	12.17
Na_2O	0.34	0.80^d	1.07^d	1.49^d	1.84	—
K_2O	0.14	0.44	0.76	0.65	0.50	—
NiO	n.d.[e]	0.07	n.d.	0.11	n.d.	—
TOTAL	100.00	98.3	98.4	98.8	99.87[f]	100.00
$\dfrac{FeO}{MgO + FeO}$	0.213	0.522	0.553	0.511	0.532	—
Q	—	2.02	3.97	—	—	—
or	0.83	2.60	4.49	3.84	2.95	—
ab	2.88	6.77	9.05	12.6	15.57	—
an	5.70	29.0	31.4	25.0	27.62	32.44
di	8.54	27.1	19.8	28.3	24.56	21.75
hy	12.59	28.2	26.8	21.5	24.17	26.35
ol	68.54	—	—	4.96	2.47	19.46
cr	0.43	0.65	0.46	0.56	—	—
il	0.49	2.03	2.36	2.09	1.92	—
	100.00	98.37	98.33	98.95	99.26	100.00

[a]Thaba Putsoa, Northern Lesotho (Boyd and Nixon, 1973), specimen no. 1611. Recalculated on anhydrous basis.
[b]Muskox intrusion (Irvine, 1970).
[c]Total iron as FeO.
[d]Part of Na lost during analysis.
[e]n.d. = not determined.
[f]Including P_2O_5, H_2O^z.

to form the protoearth, do not yield liquids with appropriate iron–magnesium ratios. The bulk sample of the unusual carbonaceous chondrite from Allende was melted 25 percent at an oxygen fugacity between that of the iron–wüstite and wüstite–magnetite buffers by Seitz and Kushiro (1974, p. 955), and the melt was analyzed. The melt is similar to some lunar ferrobasalts and has FeO/(FeO + MgO) equal to 0.797. It is not sufficiently magnesian for the formation of common terrestrial basalts; at a lower oxygen fugacity, however, the liquid should be more magnesian because of the precipitation of a metallic phase. Wager (1958, p. 38), after comparing the average analyses of 94 chondrites and a weighted average of 4 peridotites, concluded on the same basis that chondritic meteorites were not suitable mantle material. The writer believes that this conclusion should be considered as tentative until the effects of P_{0_2} are well established.

PHASE EQUILIBRIA SUPPORT FOR GARNET PERIDOTITE SOURCE

Some support is gained from relevant phase diagrams for the conclusion that on partial melting garnet peridotite will indeed yield a basaltic composition. Only one example will be given here (Figure 2-14); however, others are presented in Chapter 6. The anhydrous system

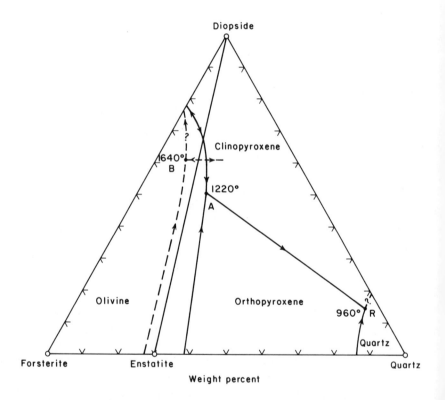

FIGURE 2-14 The system forsterite–diopside–quartz at 20 kbar under anhydrous conditions (dashed lines) and under excess H_2O conditions (solid lines). The anhydrous invariant point Opx + Cpx + Qz + L is not shown. Adapted from Kushiro (1969b, p. 282–283, Figures 6 and 7). [The same system at 10 kbar was given by Warner (1973, p. 938, Figure 4).] Invariant points A, B, and R, respectively, are analogous to andesite, basalt, and rhyolite. (With permission of the *American Journal of Science*.)

diopside–forsterite–silica displays two invariant points at 20 kbar (Kushiro, 1969b). The system contains three of the phases pertinent to garnet peridotites, and the eutectic* composition, *B*, is representative of an olivine–hypersthene normative basalt. (The second invariant point, involving orthopyroxene + clinopyroxene + quartz + liquid, has not been determined experimentally at 20 kbar and therefore does not appear in Figure 2-14.)

The same system at 20-kbar water pressure also exhibits two invariant points. The invariant point *A*, applicable to the phases pertinent to garnet peridotite, has a normative composition appropriate for andesite. [The principal arguments relating to the production of basalt under anhydrous conditions and andesite under hydrous conditions from peridotite were outlined by Yoder (1969).] For some compositions continued melting above the temperature of *A* would yield liquids on the olivine–clinopyroxene boundary curve that would be analogous to basalt. The important observations are that the entire range of compositions in diopside–enstatite–forsterite will produce, on melting, initial liquids of basaltic composition under anhydrous conditions, whereas more extensive melting is needed to produce liquids of basaltic composition under hydrous conditions.

SOURCE OF POTASSIUM?

One of the several deficiencies of the simple model of garnet peridotite as represented by forsterite + enstatite + diopside + garnet is the absence of a K_2O-bearing phase. The most likely candidate for the K_2O-bearing phase is phlogopite because of its wide range of stability and its presence in some nodules believed to come from the upper mantle. Some investigators (e.g., Oxburgh, 1964, p. 6) also suggest amphibole as a source of potassium in the mantle on the basis of the observation of O'Hara (1961, p. 251, sample X282) that almost all the potassium in a *metamorphosed* peridotite gneiss was contained in an amphibole!

---

*The temperature of the eutectic *B* is 1640 ± 10°C, and the temperature of the beginning of coprecipitation on the diopside–forsterite join of diopside solid solution and forsterite solid solution is 1635 ± 10°C at 20 kbar. Although these two temperatures are within the limits of error of each other, a maximum temperature must occur on the boundary curve between the two points. The maximum temperature marks the intersection of the boundary curve with a line connecting forsterite solid solution with the composition of the diopside solid solution having the highest thermal stability in equilibrium with olivine. That diopside solid solution will not necessarily lie on the diopside–enstatite join.

Although there is some debate about the primary or secondary nature of phlogopite, the latter would imply a still deeper source for potassium. There is greater concern about the actual amount of phlogopite present because of the heat-producing aspects of the potassium and the water content of the mica. If phlogopite is to make the major contribution of potassium to the partial melt, then there must be a sufficient amount of phlogopite to persist as a phase throughout the melting interval and successive melt withdrawals. Unfortunately, an amount of phlogopite of the order of 1 percent is considered to be excessive with regard to radiogenic heat production, based on the K_2O content of chondrites (Hurley, 1968a,b) as a limit.*

In Figure 2-15 are displayed the deduced liquidus relations among phlogopite, forsterite, and enstatite. The liquid at the beginning of melting of this assemblage, A, is greatly enriched in potassium, and phlogopite would be the first phase to be consumed.† On the other hand, the low K_2O content of oceanic basalt may reflect the absence or negligible amounts of phlogopite in the parental material (i.e., a depleted peridotite) and suggests a source of potassium from within one of the major phases (e.g., clinopyroxene).

SEISMIC VELOCITIES OF GARNET PERIDOTITE SUITABLE?

If garnet peridotite is the source rock for basaltic magmas, then the physical properties of the rock should be consistent with those properties deduced for the mantle of the earth. Adams and Williamson (1925, p. 245, footnote 9) related the seismic velocities of the primary and secondary waves to the compressibility and density of the rocks:

$$K_S/\rho = V_p{}^2 - 4/3\ V_s{}^2,$$

where K_S = the adiabatic bulk modulus, that is, the inverse of the compressibility, $1/\beta$; ρ = density; V_p = velocity of the primary wave; and V_s = velocity of the secondary wave. It appears that the seismic velocities measured in the field could give a restrictive set of solutions to the kinds of phases and their proportions in the rocks if the

---

*The minimum K_2O content of the mantle was 0.017 percent if the mantle supplied the ^{40}Ar of the atmosphere (Birch, 1951).

†According to the determined phase relations in the $Fo–Qz–An_{50}Ab_{50}–H_2O$ system with 10 percent $KAlSi_3O_8$ at 15 kbar (Kushiro, 1974, p. 245, Figure 22), amphibole, a possible alternative source of potassium, would also be consumed first at a reaction point involving forsterite, enstatite, amphibole, liquid, and vapor.

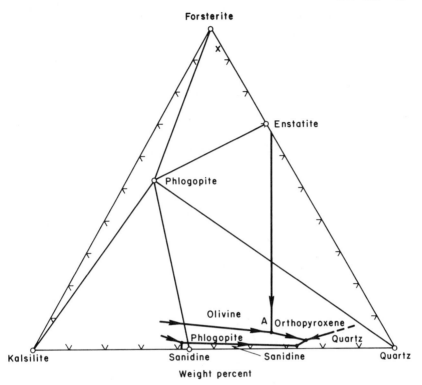

FIGURE 2-15   Projection of the system forsterite–kalsilite–quartz at 20 kbar under excess H_2O. Invariant points are exaggerated in forsterite content: They probably lie within a few percent of the baseline. Deduced from the relations presented by Luth (1967, p. 397, Figure 6f) for 3 kbar. (Leucite is not stable at 20 kbar.) The assemblage forsterite + enstatite + phlogopite begins to melt at the temperature of A, about 1130°C (Modreski and Boettcher, 1972, p. 859, Figure 2), and the liquid has the composition of A. The bulk composition X (near the forsterite–enstatite join) is analogous to that of a portion of mica peridotite. (With permission of the *American Journal of Science*.)

compressibility and density of the rocks and minerals were measured in the laboratory.

To illustrate the application of the Adams and Williamson relationship, selected values of V_p and ρ of the principal minerals of garnet peridotite are compared with the range of the same parameters believed to exist in the mantle (Figure 2-16). These carefully selected values appear to be reasonable until the ranges of these properties in eclogites with different Fe/Mg and the extent of alteration are examined. Birch (1970) clearly illustrated how sensitive the bimineralic

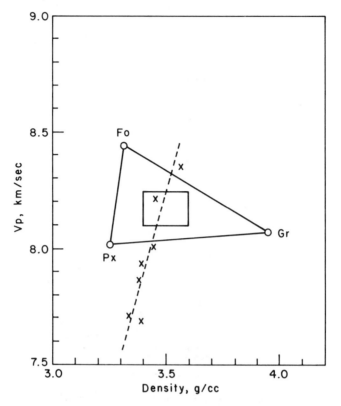

FIGURE 2-16 Selected values of primary wave velocity, V_p, and density for the major minerals in garnet peridotite (open circles), range of those values in the upper mantle (rectangle), and measured values for various eclogites ($\times$) given by Birch (1970). Iron content of eclogites increases with density. Both orthopyroxene and clinopyroxene are represented by Px. Fo = forsterite; Gr = garnet.

rock eclogite is to Fe/Mg (Figure 2-16); the denser rock is the more iron-rich. Eclogite, which should lie on the join clinopyroxene–garnet, appropriate for the mantle, would have to be iron-rich. Other published values for pyroxenite and harzburgite are too variable for a reasonable analysis. One investigator, Anderson (1970, p. 89), on the basis of his selected values, estimated that the mantle should consist of 45–75 percent olivine, 25–50 percent pyroxene, and less than 15 percent garnet. Garnet peridotite and some eclogites are possible rock types in the upper mantle; however, either the seismic data are too ambiguous for a definitive conclusion or undepleted and unaltered rocks have not yet been studied.

MONTE CARLO TEST OF PHYSICAL PARAMETERS

On the basis of parameters including the two elastic velocities, density, periods of the earth's free oscillations, dispersion of surface waves, and mass and moment of inertia of the earth, Press (1968, 1970) made a random test by a Monte Carlo method using successive constraints to produce 18 possible models of the suboceanic upper mantle. He found (Figure 2-17) that an eclogite facies model (~50 percent eclogite) was generally acceptable between 80 and 150 km with a transition zone to 300 km, below which a peridotite model was favored. Press believed

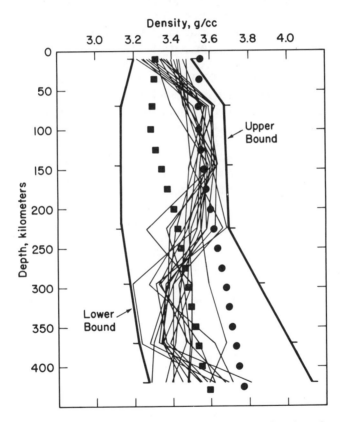

FIGURE 2-17 Various successful density models for the sub-oceanic upper mantle (from Press, 1969, p. 174, Figure 1). Bounds define range assigned in the Monte Carlo selection. Dots are density values for the eclogite model, and squares are those for the pyrolite model of Clark and Ringwood (1964). (With permission. Copyrighted by the American Association for the Advancement of Science.)

that "eclogite" is the high-pressure crystalline product derived by the partial melting (15 percent) of peridotite that has not been wholly depleted of eclogitic (basaltic) components. His model is gravitationally unstable, and he emphasized that the considerable latitude in properties may not necessarily be due to heterogeneity in the mantle. For example, density contrasts may be stabilized by thermal gradients.

COMPOSITION(S) OF THE MANTLE NOT DETERMINED

What, then, is the composition of the mantle? Because of the lack of precise information on the kind, abundance, and composition of the minerals, the bulk composition of the mantle from place to place cannot be specified. The variation of Pb and Sr isotopes in basaltic magmas is at present interpreted as resulting from the melting of masses of different composition or age (Gast *et al.*, 1964; Hofmann and Hart, 1975; Brooks *et al.*, 1976). According to Erlank and Kable (1976, p. 287), element abundance ratios such as K/Rb, K/Ba, and Zr/Nb also lead to the conclusion that the chemical differences between some basalts cannot be attributed to fractionation and that different source regions are involved.

The existence of large masses in the mantle unique in age or composition implies that the mantle endured several remeltings, perhaps early in the earth's history when the heat flow was several times greater than at present (see Figure 5-8). The rise of diapiric masses, for example, may have then been commonplace, and the ascent of some of these masses may have terminated at moderate depths when they encountered the base of a depleted, less dense layer. These arrested and cooled masses, now nearer the surface, may be the source of magma released on a later cycle of reheating (see Chapter 10). A discussion of the heterogeneity of the mantle is deferred until the principles of the melting process are examined (see Chapter 6).

GARNET PERIDOTITE ACCEPTED AS SOURCE—TENTATIVELY

On the basis of the above arguments, it is concluded that garnet peridotite is the preferred source of basaltic liquids because:

1. The occurrences of garnet peridotite are appropriate to deep-seated environments.
2. The close compositional relationship to meteorites supports its derivation from material accumulated in the primordial earth.

3. Partial melts of natural samples at high pressure have basaltic compositions.

4. The mineral assemblage was found experimentally to be stable at high pressures and temperatures.

5. It has appropriate densities and seismic velocities.

For these reasons most of the following arguments on magma genera-tion are based on the minerals in this rock type. The reader should keep in mind that this is a working conclusion, not well established, and it will be tested in succeeding sections.

To give the reader a physical image of the kind of material envisaged in the mantle, a hand specimen of garnet peridotite is displayed in Plate 1 (p. 44a). The appearance of the rock has given rise to the "plum-pudding" concept of the mantle, where garnet and clinopyroxene, the potential basalt "plums," sit in a "pudding" of olivine and orthopy-roxene. It is likely that under the pressure and temperature conditions of the mantle the rock where stable would not exhibit the beautiful colors displayed in the plate but would be nearly white hot.

3 Depth of Melting

As regards the *source of the magmas*, it is clear that a fundamental inquiry would take us back into a region of speculation from which we had already turned away as unprofitable. It does not appear, however, that ignorance concerning the condition of the interior of the Earth as a whole need materially embarrass the pursuit of the object immediately before us.

Harker (1909, p. 33)

Much of the debate about where melting takes place in the mantle rests heavily on notions of the origin of the earth. Because the composition of the earth and the distribution of rock types with depth depend on the nature of the earth's formation, it is critical to know (a) whether the earth accumulated relatively cool and was heated up by the decay of radioactive elements and by conversion of gravitational energy to thermal energy or (b) whether it was initially molten and differentiated with cooling.

ORIGIN OF PROTOEARTH: COOL ACCUMULATION, DIFFERENTIATED MELT, OR CONTINUUM OF BOTH?

These major concepts can perhaps be best described with a schematic two-component system. Although the triple point involving liquid for any simple igneous rock forming silicate has not yet been determined in the laboratory, the relations are probably close to those displayed in

44

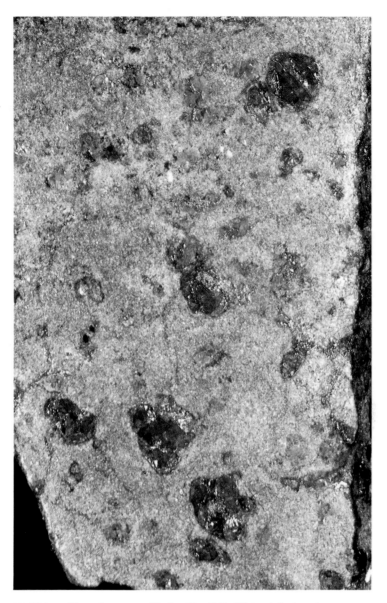

PLATE 1. Natural garnet peridotite, No. 1242. Olivine (olive), orthopyroxene (gray), clinopyroxene (bright green), garnet (burgundy), Sunnmøre, Norway. Width of polished section is 7 cm. Specimen courtesy of Dr. H. H. Schmitt. Photograph by Dr. P. E. Hare.

44a

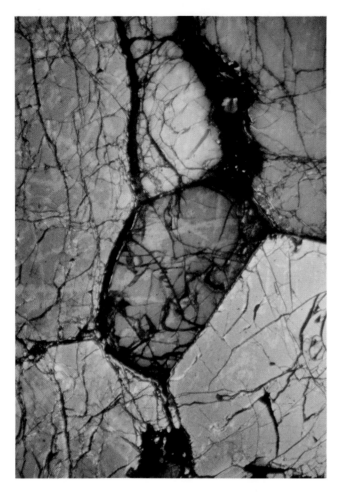

PLATE 2. Granular garnet peridotite, No. 1859N. Monastery Mine, Republic of South Africa. Width of section is 8 mm. Photograph courtesy of Dr. F. R. Boyd.

44b

Figure 3-1*A* for forsterite–H_2.* If the effective pressure during the condensation of the earth was about 10^{-2} bar, the earth may have passed through a molten stage, depending on the bulk composition. For example, composition *X* in Figure 3-1*B* would, on cooling, pass through a liquid region. On the other hand, a composition such as *Y*, highly dilute with respect to potential silicates, would not encounter liquid on cooling. If the effective pressure on aggregation was approximately 10^{-6} bar (Figure 3-1*C*), the gas would have condensed directly into crystalline material without encountering a liquid phase for any composition. The eutectic, not shown in Figure 3-1*C*, would probably lie at a temperature of the order of $-260°C$. A most useful set of practical experiments to be performed would no doubt include the system forsterite–hydrogen or diopside–hydrogen at appropriate pressures as an analogue of the condensation of a star. Identification of the species in the gas phase would be very helpful to those using thermodynamic data to calculate the phase relations.

The two alternatives displayed in Figure 3-1*A* are not necessarily the extreme pressure limits, and, in fact, a range of pressures and compositions probably was effective during the cooling and aggregation of the protoearth. The relationship of the rate of aggregation to the cooling rate may have had considerable influence on the texture of the accumulated material. For example, if the gas cloud cooled first at 10^{-2} bar (Figure 3-1*B*, composition *X*) to liquid droplets, which then crystallized before aggregation, the texture might be similar to that of a chondrite. The chondrules could have formed by the rapid cooling or quenching of the droplets. On the other hand, if the gas cloud cooled directly to crystalline aggregates (Figure 3-1*B*, composition *Y*; or Figure 3-1*C*, composition *Z*), the texture might be akin to that of an achondrite.

Another factor of considerable importance is the time required to collect or sweep up the condensing materials. If the sweep-up time was short, the individual masses would be small and the heating on impact would be preserved. On the other hand, if many orbits were necessary for aggregation, large masses were assembled, and the accumulation was slow, the heating on impact might be dissipated. In the first

---

*For another analysis of the problem wherein solar abundances are taken into consideration and point *A* is believed to lie at several hundred bars, the reader should consult Wood (1963, p. 156, Figure 3). Because the planets in the solar system are probably of different compositions, the present solar abundances do not appear to be appropriate for the initial composition of the specific gaseous region from which the earth was formed. Thermodynamic calculations of the "triple" point depend on knowledge of the gaseous species; however, the species in gases of suitable compositions existing prior to the formation of a protoearth have not been ascertained experimentally.

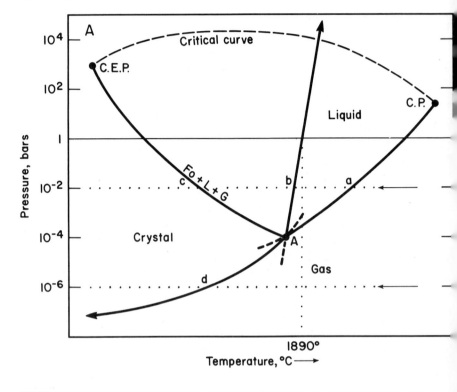

instance, a relatively homogeneous protoearth would be formed, whereas in the second instance, the protoearth formed would be heterogeneous to a degree dependent on the size of the aggregated masses prior to impact.

CRUST FORMATION THROUGH VOLCANISM

It will be assumed for the present purposes that the pressure through-out the condensing toroidal cloud was indeed low and that the earth mainly accumulated from a relatively cool gas directly into crystals without an intervening molten stage except at the center of the toroid. Birch (1965) thought that conversion of the gravitational potential energy on core formation would raise the temperature of the mass 1600°C on the average; however, the efficiency of the conversion might have been low and much of the heat might have dissipated so that only

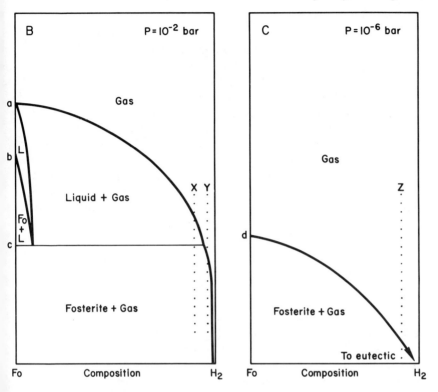

FIGURE 3-1 *A*. Schematic pressure–temperature diagram for the high-temperature portion of the forsterite–H_2 system, assuming congruent relations on all univariant curves. *A* = the invariant point for forsterite; *C.P.* = critical point for forsterite; *C.E.P.* = first critical end point of the binary system. Lowercase letters refer to points occurring in the constant pressure sections (marked with light arrows) displayed in Figures 3-1*B* and 3-1*C*. Heavy arrows indicate that the curves continue. Heavy dashed curves about *A* are metastable extensions of equilibrium curves (heavy lines) for phases of forsterite composition.

B. Schematic temperature–composition diagram for the high-temperature portion of the forsterite–H_2 system at a total pressure equal to 10^{-2} bar. Lowercase lettered points are identified in Figure 3-1*A*. *X* and *Y*, respectively, are compositions that illustrate possible formation of chondritic and achondritic textures on cooling.

C. Schematic temperature–composition diagram for the high-temperature portion of the forsterite–H_2 system at a total pressure equal to 10^{-6} bar. Lowercase lettered point is identified in Figure 3-1*A*. *Z* is a composition illustrating the cooling behavior of an earth-forming gas cloud at very low pressure and relatively low temperature.

the core and a portion of the lower mantle remained or became molten. The ambient temperature on initiation of condensation of the principal phases in plagioclase peridotite would be well below 1250°C. That temperature is the eutectic for anorthite + forsterite + diopside + enstatite at 1 atm, according to Schairer and Yoder (1970), p. 212, Figure 11), and the vaporus eutectic at 10^{-6} bar, for example, would be at much lower temperatures. The crust, formed at the earliest stage by volcanic processes, was probably andesitic in composition because of the degassing of the mantle, a process that is believed to have led to the formation of the oceans. Andesite is thought to have been derived by the hydrous partial melting of the mantle (see invariant point A, Figure 2-14), according to the views expressed by Yoder (1969). Water presumably was trapped in hydrous minerals, as a free phase, or was the oxidation product of trapped primordial hydrogen, the atmosphere in which the earth aggregated.

IMPACTING, DEEP SURFICIAL TURNOVER, AND PARTIAL MELTING

Crust formation implies that there is a region at some depth below the surface depleted of crustal materials. On the basis of present knowledge of lunar history it is likely that the earth endured surficial melting to a depth of the order of 50 km as a result of the great impacting that is believed to have culminated at about 4 b.y. (Tera et al., 1974). Whereas the average depth of melting may have been about 50 km, a few areas may have suffered impact melting to a depth as great as 200 km, comparable to the radius of the largest of impacting bodies. The question arises as to whether sufficient time has elapsed for all the material melted on impact to have crystallized. In other words, are there pockets of molten material residual from the impacting, which must have been more severe than that observed on the lunar surface?

The turnover of crust and upper mantle materials was presumably very thorough as a result of impacting. It should be recalled that the oldest rocks already contain quartz-bearing sediments and are enclosed by granitic* gneisses, about 3.75 b.y. old (Moorbath et al., 1972). The quartz was most likely derived from the churned andesitic (average 15 percent modal quartz) crust. It is indeed remarkable that there is no evidence yet found on earth of impactite equivalent to the lunar regolith. If plate tectonic processes were operating during, as well as

*Granitic melts are not a possible partial melting product of peridotite, but are presumed to be (a) a derivative of basaltic or andesitic melts separated from their parental material or (b) the partial melting product of sedimentary materials. The origin of rhyolites and granites is in itself a story worthy of a separate volume.

after, Precambrian times, the crust might have been recycled several times. Erosion must have been very efficient and crustal subduction complete to remove all traces.

MAGMA—A RESIDUUM FROM EARTH FORMATION?

There are obviously too many options available to provide a unique sequence of events describing the origin of the earth from the few facts known. The reader should keep in mind the possible existence of residual pockets of magma, at least in early Precambrian time, resulting from (a) an initially molten earth, (b) an earth reheated by gravitational energy conversion, (c) a deeply impacted surface, and (d) short-lived radioactive isotope decay. The dimensions, of undefined aspect ratio,* of these pockets, if existent today, are presumed to be less than the wavelength of seismic body waves now received at the surface, that is, several tens of kilometers (Shimozuru, 1963, p. 192). The cooling rate, dependent on the temperature of the ambient rocks, of such small pockets was probably sufficient that those residual magmas became crystalline by the end of the Archean. Of course, there is no way of knowing what the original dimensions of trapped magmas were.

SEISMIC EVIDENCE—AUXILIARY MAGMA CHAMBERS

One must depend mainly on seismic evidence for indications of the depth of generation of contemporary melts. Magnetic and electrical resistivity measurements are not sufficiently specific to delineate a magma body, but they do provide useful confirmatory evidence. Broad areas of relatively high temperature have been deduced by Porath *et al.* (1970, pp. 258–259) by correlation with high electrical conductivity in the mantle and by Morgan (1972, p. 206, Figure 4) by correlation with high isostatic gravity anomalies. Intuitively, one would expect the thermal expansion resulting from near solidus temperatures to produce low gravity; however, the upward flow of denser, as well as hotter, mantle material into the top of the mantle and crust may produce a net increase in gravity.

In general, the crust and mantle transmit both primary and secondary seismic waves at all depths, and therefore those regions are considered to be predominantly solid. This view is supported by data on tidal deformation and the fundamental oscillations of the earth. Changes in the velocity and attenuation of seismic waves in certain regions,

---

*The aspect ratio is the ratio of minor to major axes of an ellipsoidal inclusion. The aspect ratio of liquid inclusions greatly influences the rigidity and attenuation in the host rock (Walsh, 1969).

however, suggest that some areas may not be completely solid. For example, Kubota and Berg (1967), after the studies of Gorshkov (1958), succeeded in outlining areas in the Katmai region, Alaska, where the shear wave was attenuated, presumably by liquid. They outlined shallow magma reservoirs using four seismic stations. By observing a large number of earthquakes, Kubota and Berg noted ray paths along which seismic stations did not receive the shear wave from specific earthquake centers. The "screening effect" led to the deduction of the boundaries of the magma chamber. The vertical distribution of the chambers deduced is shown in Figure 3-2. These chambers are considered by the writer to be auxiliary (transported magmas) and not primary (site of generation).

Gorshkov (1958) concluded that the chamber under Kliuchevsky volcano, Kamchatka, was between 50 and 75 km deep, and he believed that its shape was that of a convex lens or perhaps a triaxial ellipsoid. More recent studies by Fedotov and Tokarev (1974) indicated "that the roof of the zone of intense fractional melting and magma generation" is at a depth of 50–60 km; however, the "main zone of magma generation is probably located within the depth interval from 120 to 200–250 km. . . ." Eaton *et al.* (1975) outlined a partially molten region beneath Yellowstone National Park, Wyoming, extending from near the surface to about 100 km, on the basis of a 10 percent reduction in *P*-wave velocity. The diameter of the region showing seismic wave attenuation is about 30 km near the surface, expanding to about 60 km at 60 km

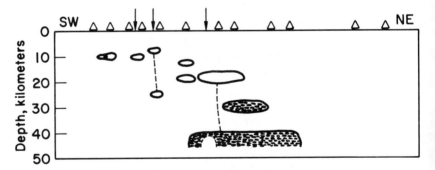

FIGURE 3-2   Vertical distribution of magma chambers (enclosed areas) and possible conduits (dashed lines) along the Katmai, Alaska, volcanic range according to Kubota and Berg (1967, p. 202, Figure 16). The two deepest chambers outlined may consist of many smaller molten pockets, according to an alternative suggestion of Shimozuru (1963). Volcanoes (triangles) and seismic stations (arrows) are noted. The fourth seismic station is at Kodiak, about 150 km to the southeast of the range. Vertical and horizontal scales are about equal. (With permission of the *Bulletin Volcanologique*.)

depth. On the basis of teleseismic *P*-wave arrival time residuals, Steeples and Iyer (1976) concluded that anomalously hot rock, or possibly partially molten rock, lies at a depth of more than 7 km and less than 40 km beneath the Long Valley, California, caldera. The velocity in contrast with the surrounding material is lower by more than 5 percent and is probably in the range 10–15 percent. Although the areal extent is poorly constrained, a large amount of detailed geological and geophysical evidence supports their hypothesis of a residual auxiliary magma chamber. Sanford *et al.* (1973) interpret the large amplitudes of reflected shear waves (both S_rP and S_rS) from a surface at a depth of about 18 km under Socorro, New Mexico, as due to the presence of "hot material of low rigidity." The region is known for its high heat flow, hot springs, and recent basaltic eruptions.

LOW-VELOCITY ZONE—EVIDENCE OF PARTIAL MELTING?

If the Alaskan and Kamchatka chambers are auxiliary and not the site of generation, then it is necessary to look at broader-scale features, unfortunately with great loss of resolution. The increase in data from the study of surface waves in the past 10 years, however, has greatly improved the understanding of the deeper structures. The change of seismic velocities with depth according to Nur (1974, p. 302) is given in Figure 3-3. The region of diminished velocities extends from about 70 to 150 km. In the low-velocity zone, discovered by Gutenberg (1926), the amplitude, a parameter difficult to measure with precision, decreases strongly (Figure 3-4). The profiles given by Nur are not necessarily worldwide, and some authors believe the low-velocity zone is absent over large areas of the oceans and continental shields (Knopoff, 1972; Jordan, 1975). Some data on vertical shear-wave and Love-wave dispersion indicate a fundamental difference between the mantle under oceans and continents: that debate is still in progress. The velocity reversal, where observed, may be due to (a) a high-temperature gradient, (b) a solid-phase change, (c) a compositional change, or (d) partial melting.

The constraints of heat flow and potential phase changes observed in the laboratory support the first two possibilities. A change of olivine composition from Fo_{90} to Fo_{70} could account for the velocity change, according to Birch (1969, p. 34). The 5 percent drop in velocity could imply the presence of about 7 percent liquid if its $V_p \approx 2.5$ km/sec. Machado (1974, p. 258–259) noted that the observed rigidity of the low-velocity zone can be accounted for if the value for crystalline

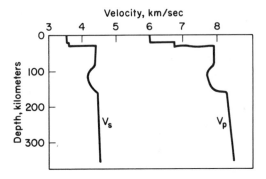

FIGURE 3-3 Approximate velocities of the primary, V_p, and secondary, V_s, waves in the upper mantle, according to Nur (1974, p. 302, Figure 4.10).

peridotite is reduced by the presence of 5 percent fluid inclusions. Kubota and Berg (1967, p. 212) explained the increase in Poisson's ratio for the region under Katmai by assuming that a minimum of 8 percent of the volume of the upper mantle is molten.

PARTIALLY MELTED ZONE—EXPERIMENTAL EVIDENCE

The interpretation of the low-velocity zone as a region of partial melting is supported by the experiments of Spetzler and Anderson (1968) and Anderson and Spetzler (1970). They determined the velocity of both longitudinal and shear waves in cylinders of ice containing irregular pockets of NaCl brine. A large drop in velocity was observed with very small amounts of melt present. They also calculated the influence of the aspect ratio of the droplets of melt on the velocity (Figure 3-5) according to the Eshelby–Walsh theory (Eshelby, 1957; Walsh, 1969).

By using vitreous silica rods to transmit the compressional waves through melt to a transducer, Murase and Suzuki (1966) and later Murase and McBirney (1973, p. 3572) obtained a value of 2–3 km/sec in the temperature range of 1100–1500°C for molten basalts. In both studies an abrupt drop in velocity was observed as melting began.* It is evident from Figure 3-5 that thin, intergranular films of a very small amount of melt could account for the low-velocity zone. Birch (1969), using olivine and basaltic glass inclusions as a model, found that

---

*There are many opportunities for experimental work in seismology. The present need appears to be for a high-temperature transducer to measure velocities and attenuation in silicate melts at appropriate mantle conditions. Progress has already been made with some ceramic electronic components capable of withstanding 500°C (J. R. Rowley, personal communication, 1975).

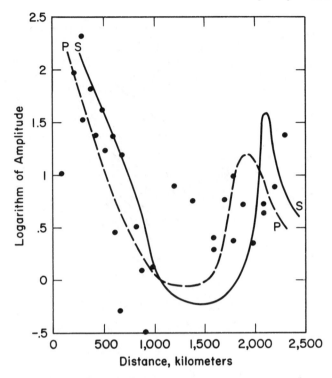

FIGURE 3-4 Worldwide average seismic amplitudes for primary, *P*, and secondary, *S*, waves at various recording distances, according to Gutenberg (1959). Amplitude variation supported by primary waves from nuclear explosions (solid dots) originating in New Mexico and Nevada (Anderson, 1962, p. 55). Drop in amplitude from 100 to 1,000 km is attributed to the low-velocity zone. (With permission of Scientific American, Inc.)

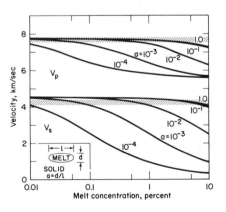

FIGURE 3-5 Variation of velocity of primary (compressional) wave, V_p, and secondary (shear) wave, V_s, at 1200°C with melt concentration assuming different aspect ratios, *a*, as defined in the inset figure, of droplets, according to Anderson and Spetzler (1970, p. 63, Figure 2). The major and minor axes of the ellipsoidal droplets are *L* and *d*, respectively. Ruled region is the low-velocity zone. (With permission of the Elsevier Scientific Publishing Company.)

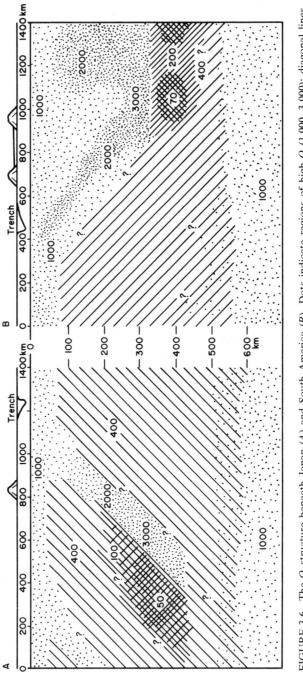

FIGURE 3-6 The Q structure beneath Japan (A) and South America (B). Dots indicate regions of high Q (1,000–3,000); diagonal lines, intermediate Q values (300–500); and cross-hatching, low Q values (50–100). After Sacks and Okada (1974, p. 215, Figure 5). (With permission of the Elsevier Scientific Publishing Company.)

54

spherical droplets and 6 percent melt could explain the low-velocity zone.

Because $V_s^2 = \mu/\rho$, where μ is the shear modulus (or rigidity), a density increase could also account for the drop in shear velocity. Press (1969) has already shown (Figure 2-17) that density increases in the same zone; the density change would not, however, account for the threefold increase in attenuation (see below) according to Nur (1974, p. 303).

ATTENUATION OF SEISMIC WAVES

The attenuation, damping, or dissipative function of seismic waves is usually expressed in terms of a dimensionless quantitity, Q, defined as 2π times the ratio of the stored energy to the dissipated energy per cycle. It is related in an unknown way to the viscosity. If the viscosity is low, it is likely that the region is hot and possibly partially melted. Sacks and Okada (1974) determined Q for sections across Japan and the Chile–Peru region of South America (Figure 3-6). If Q values of 50 to 100 imply melting, then the source of magma under Japan is at depths of 300–400 km, whereas the source under Chile–Peru is at about 400 km. The inferred melt region under Japan is above a subducting plate characterized by very high Q values (1,000–3,000), whereas under Chile–Peru the subducting plate abruptly terminates at the inferred melt region. Sacks and Okada noted the absence of low Q values in the vicinity of known active volcanoes and suggested that "there must be a narrow pipe transporting the magma from depths." The aseismicity in the 350–500-km region beneath Chile–Peru supports the view that the temperature is so high that the rocks cannot store strain energy and are unable to fracture elastically. Earthquakes are known to occur at depths as great as 700 km.

SUMMARY

The seismic evidence indicates that magma may be generated in some broad regions extending generally from 50 to 170 km in depth. In other regions restricted to specific structural environments, the source of magma may be as deep as 300–400 km. The seismic data are insufficient for independent determination of the shape of the chamber, the distribution of the liquid, and the extent of melting. It is difficult to distinguish whether the deduced magma chamber is the site of generation or merely the region in which liquid is separated from a transported, partially molten mush. Much of the debate relating magma generation to tectonics will probably falter on this point in the absence of distinguishing criteria. Many geophysicists prefer a model of a magma chamber that is an oblate spheroid, elongated horizontally, to represent the partially molten chamber containing up to 8 percent liquid in droplets ranging in shape from thin films to spheres. (The form of the dispersed liquid will be discussed further in Chapter 9.) The field geologist, whose opinions are based mainly on investigations of cooled and eroded chambers of transported magma, prefers models of magma chambers that are more irregular in shape with the base undefined, a lenticular body, or an extended layer.

4 Melting Processes

MULTITUDE OF MECHANISMS

Now that the reader has a rough idea of *what* is melting and *where* it melts, an effort will be made to present the various ideas expressed on *how* the melt is generated. An exceptionally large number of mechanisms have been proposed, each having some advantages—and usually several disadvantages. Some mechanisms are specific to certain structural environments, whereas others are of general applicability. The reader will find it useful to consider each mechanism in relation to the various regimes of plate tectonics. The mechanisms involve changes of pressure, temperature, composition, position, and stress, as well as contrasts in various other properties. A change in practically every known parameter of consequence has been nominated as a cause of melting. Not all the mechanisms will be described; those chosen, however, will provide the framework for substituting other parameters.

MELTING BEHAVIOR OF CRITICAL MINERALS

First, it is necessary to look at the melting behavior of the four critical end-member minerals in garnet peridotite because several of the processes depend on the character of its melting curve. Historically, it was thought that the melting curve of a substance would become asymptotic to a fixed value with increasing pressure (Figure 4-1A), reach a maximum (Figure 4-1B), terminate in a critical point (Figure

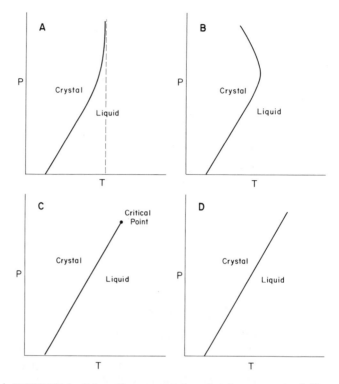

FIGURE 4-1   Schematic representation of various concepts of silicate melting with increasing pressure: A, Asymptotic approach to
limiting melting temperature. B, Melting temperature attains a
maximum. C, Melting curve terminates at a critical point. D, Continuous rise of melting temperature within observational range.

4-1C), or continue indefinitely (Figure 4-1D). These views were super-
seded as data at higher pressures became available.

Forsterite, Mg_2SiO_4 (Figure 4-2), melts with a gradient of 4.77°C/kbar
up to 50 kbar, according to Davis and England (1964, p. 1115, Figure 2).
Akimoto (1972) found that the melting curve is terminated by a phase
change to a β phase, an orthorhombic structure termed a "modified"
spinel by Morimoto et al. (1970; see also Moore and Smith, 1970), at
pressures near 125 kbar at about 1000°C and near 115 kbar at 800°C.

Diopside, $CaMgSi_2O_6$, melts (Figure 4-2) with an initial gradient of
13°C/kbar up to 5 kbar (Yoder, 1952) that decreases to 7.5°C/kbar at 50
kbar (Boyd and England, 1963, p. 321, Figure 6). At 1 atm diopside
melts incongruently to a clinopyroxene solid solution relatively rich in
$MgSiO_3$, and the liquid becomes enriched in $CaSiO_3$. The melting curve

in Figure 4-2 marks the disappearance of the last crystal. The pressures to which the incongruent melting character of diopside persists are not known. Arguing from the behavior of the germanate analogue, Ringwood (1966, p. 381, Table 5) suggested that the melting curve is probably terminated by the breakdown of diopside to garnet and ilmenite structures.

Enstatite, $MgSiO_3$, melts (Figure 4-2) with an initial gradient of 12.8°C/kbar that decreases to 6°C/kbar at about 50 kbar (Boyd *et al.*, 1964, p. 2104, Figure 1). The very-low-pressure region is complicated by incongruent melting to forsterite + liquid and polymorphism to protoenstatite, but the nature and position of the inversion curve are being debated (see Kushiro *et al.*, 1968a; Chen and Presnall, 1975). The melting curve is probably terminated by a breakdown to β-Mg_2SiO_4 and stishovite (Ringwood, 1970, p. 128) at pressures in excess of 200 kbar.

Pyrope, $Mg_3Al_2Si_3O_{12}$, melts (Figure 4-2) with a gradient of 14.1°C/kbar above 21.6 kbar and 1510°C (Boyd and England, 1959a, p. 84, Figure 1). At lower pressures pyrope is represented by a succession of

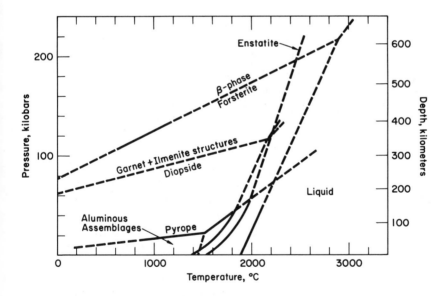

FIGURE 4-2 Melting behavior of the four major end-member minerals in garnet peridotite. The high-pressure breakdown products of diopside represent structure types and not mineral compositions. Solid lines indicate portions of curves determined experimentally. Dashed lines are extrapolations based on preliminary data, iron-rich members, or analogues as noted in text.

assemblages, which include forsterite + cordierite + spinel at 1 atm. Pyrope is the only garnet end member not stable at 1 atm, and it has the largest change in melting temperature with pressure. It is stable to very high pressures (<300 kbar), where it breaks down to a mixture of $MgSiO_3$ (perovskite structure) plus corundum (Liu, 1974).

The curves outlined in Figure 4-2 illustrate the general concept that the melting temperatures of principal silicates rise with increasing pressure and are terminated with a phase change or reaction. Evaluations of solid solution effects, especially those involving Fe, suggest that these four critical phases in garnet peridotite will probably be stable to the greatest depths now being considered as possible sites for magma generation (i.e., ~400 km). It is, however, more important to know the stability of the *assemblage* of minerals, and those limits are discussed in Chapter 6.

THE MECHANISMS

Some of the proposed mechanisms for bringing about the melting of crystalline parental material are outlined as follows.

Stress relief
  Tension (Yoder, 1952)
  Compression (Uffen, 1959)
Thermal rise to cusp in melting curve
  Compositional cusp (Buddington, 1943; Hess, 1960)
  Phase-change cusp (Presnall et al., 1973)
Convective rise
  Isentropic rise (Verhoogen, 1954)
  Isenthalpic rise (Waldbaum, 1971)
Property perturbation
  Thermal conductivity (Lubimova, 1958; McBirney, 1963)
  Density (Grout, 1945; Ramberg, 1972)
Mechanical energy conversion to heat
  Thrust faulting (Nutting, 1929; DeLury, 1944)
  Subducting plate (Toksöz et al., 1971; McKenzie and Sclater, 1968)
  Propagating crack (Griggs, 1954; Griggs and Handin, 1960)
  Regenerative feedback (Shaw, 1969; Anderson and Perkins, 1975)
  Tidal dissipation (Shaw, 1970)
Compositional change
  Diffusion under pressure differential (Bell and Mao, 1972)
  Addition of volatiles (Yoder and Tilley, 1962; Bailey, 1970)

Radioactive heat production
  Internal (Joly, 1909)
  External (Holmes, 1915)
Residual liquids from protoearth
  Primordial condensed gas (Rittmann, 1962)
  Residual liquids squeezed out on crystallization (Chamberlin and
    Salisbury, 1905)
Exothermic chemical reactions
  Surficial (Jaggar, 1917)
  In chamber (Daly, 1911; Day and Shepherd, 1913)

No measure of importance should be attached to the order of their presentation.

STRESS RELIEF

TENSION

The first melting curve of any common rock-forming silicate measured under high pressures in the laboratory gave rise to a mechanism involving relief of stress. Yoder (1952) studied the melting behavior of diopside to the then very high pressure of 5 kbar at temperatures up to 1500°C in an argon-gas, high-pressure apparatus. He believed that the pressure on the source region, then considered to be at or near the base of the crust, could be relieved, for example, by gentle arching or faulting of the overlying rock. On release of pressure, the source has a temperature in excess of that required for melting at the reduced pressure, and melting takes place. His case for a 5-kbar release is reproduced in Figure 4-3. The heat for melting was supplied from the source region itself (that is, the process is adiabatic), and about 6 percent melt is produced in the model described. The existence of a 5-kbar stress in the mantle is not now considered tenable, as discussed in Chapter 9.

COMPRESSION

The stress-release concept was embellished by Uffen (1959) and supported by Uffen and Jessop (1963), who considered the effects of compression rather than tension. Uffen's model is reproduced in Figure 4-4. Cyclical compressions and failures are required to take place in the presence of continuously rising temperature. Eventually, one failure is sufficient to reduce the pressure to that below the melting

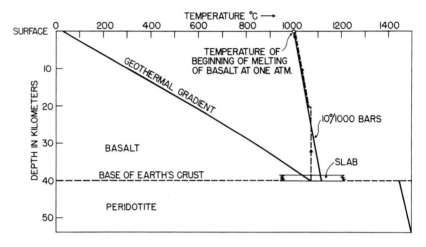

FIGURE 4-3   Magma generation model of Yoder (1952, p. 372, Figure 5), based on stress relief under tension. Dashed curve with arrows shows effective pressure release and subsequent temperature change as magma generated moves to the surface. (With permission of the University of Chicago Press.)

curve, and melting takes place. Field support for this version of the stress-release concept may be derived from the work of Fahrig and Wanless (1963). They presented a map of Canada (Figure 4-5) showing swarms of dikes parallel to the compressive fold axes of the various regions. Apparently, elastic rebound after a period of crustal compression and heating resulted in a low-pressure belt where basaltic magmas

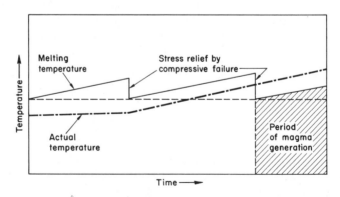

FIGURE 4-4   Magma generation model of Uffen (1959, p. 118, Figure 2) based on stress relief with a compressive failure in region of rising temperature. (With permission. Copyrighted by the American Geophysical Union.)

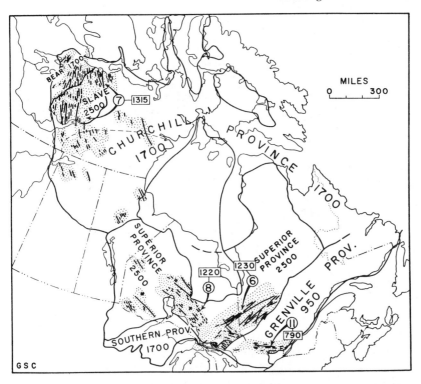

FIGURE 4-5 Diabase dike swarms of the Canadian Shield postdating compressive mountain-building stage in provinces. Map compiled by the Geological Survey of Canada (GSC), Fahrig and Wanless (1963, p. 936, Figure 2). Encircled numbers refer to localities of dikes having measured ages in millions of years recorded in associated rectangles. [With permission of Macmillan (Journals) Ltd.]

intruded. Ages determined on dikes in the swarms noted in Figure 4-5 postdate the age of the provinces in which they occur.

THERMAL RISE TO CUSP IN MELTING CURVE

COMPOSITIONAL CUSP

Hess (1960) developed the idea of Buddington (1943, p. 137) that melting could be achieved by depressing the crustal layers of the earth into a warmer region. In Hess's model, illustrated in Figure 4-6, subsequent rise of temperature in two basalt layers is sufficient to nick the melting curve, and partial melting takes place at C and D. Hess emphasized partial melting rather than complete melting and pointed

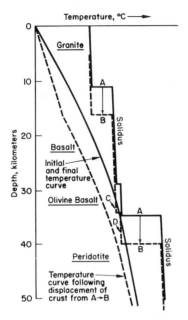

FIGURE 4-6 Compositional cusp and depression model for magma generation of Hess (1960, p. 180, Figure 36). Curves *A* and *B*, respectively, refer to the solidus of the rock types underlined before and after displacement. Re-establishment of thermal equilibrium (temperatures on dashed curve return to initial values) results in partial melting at *C* and *D*. (With permission of the Geological Society of America.)

out that the continental crust is probably more heterogeneous than a simple layered structure. In his view, the requirements for enthalpy of melting merely delay the rise of temperature. It is evident that in this model a considerable amount of downwarping would be required to initiate melting. As in the stress-release hypotheses, close proximity of the geothermal gradient curve to the melting curve is necessary.

PHASE-CHANGE CUSP

Whereas Hess appealed to appropriate changes in bulk composition to bring about interception of the geothermal gradient with the melting curve, Presnall *et al.* (1973) believed that the cusps arising in the solidus of a single bulk composition, because of change in phase under anhydrous conditions, are more likely conditions for melting to begin. With increasing temperature, melting will occur first at the lowest temperature on the solidus, which they indicated is at depths where a change of phase occurs. The changes in slope of the solidus for the anorthite + forsterite reactions at about 1335°C and 8.7 kbar as well as 1410°C and 17.2 kbar, shown in Figure 2-8, would be likely analogous points for the temperature on a rising geotherm to meet that of the solidus. Strong support for liquid generation at such invariant points is given in Chapter 7.

CONVECTIVE RISE

ISENTROPIC RISE

Instead of *depressing* the source material into a region of appropriate temperatures for melting, Verhoogen (1954) suggested raising the source material by solid-state convection (Figure 4-7) into a region appropriate for melting. Because of the large vertical and horizontal transport of heat by mass movements, Verhoogen (1973) did not accept the concept of a geotherm based on a conductive heat model. He argued that a temperature gradient less than a few degrees per kilometer would induce convection. Material starting at S rises on an adiabat (0.2° to 0.5°C/km), and melting begins at M, the solidus of the material. As melting continues, heat is absorbed in the process and the mass cools until M_1 is reached, where liquid segregates, according to Verhoogen. The liquid rises to the surface and has a temperature of M_2. Neither the degree of melting achieved nor the cause of segregation of liquid is specified. The heat requirements, density contrasts, and minimum depth of origin are discussed in Chapter 5.

A convecting system is a delicate balance between convective heat transfer and conductive heat losses through the boundaries of the mass. Shaw (1965, p. 145) considered such a system a "transient condition." He pointed out that when heat transfer is greater than heat losses, heating of the marginal material occurs, the temperature gradient diminishes, and convection dies out. Alternatively, if the heat transfer is less than the heat losses, the magma cools, viscosity increases, crystallization begins, the temperature gradient diminishes, and convection dies out. Quantitative analysis of the problem would be a most difficult, but rewarding, task. The adiabatic rise of material is commonly assumed in magma generation models; the assumption should be considered as tenuous.

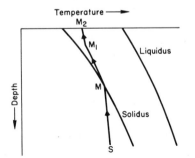

FIGURE 4-7 Magma generation model of Verhoogen (1954; 1973, p. 516, Figure 1) based on the convective rise of matter from S along an adiabat that intersects the solidus, where melting begins at M. Liquid segregation takes place at M_1 and rises on another adiabat to the surface, where the liquid has the temperature M_2. (With permission of the Geological Society of America.)

ISENTHALPIC RISE

Waldbaum (1971) suggested an isenthalpic rather than an isentropic rise of a mass to achieve melting conditions. The irreversible adiabatic decompression is given by the Joule–Thomson (1854) equation

$$\left.\frac{dT}{dP}\right|_{H} = \frac{V(T\alpha_P - 1)}{C_P}$$

where V = volume, α_P = isobaric thermal expansion, C_P = isobaric heat capacity, and H = enthalpy. The isenthalpic adiabatic rise for a garnet peridotite would be about $-23°C/kbar$, according to Waldbaum's mineral values, compared with about $0.5°C/kbar$ for an isentropic adiabatic rise. Ramberg (1971) and Shaw and Jackson (1973, p. 8648–8649) pointed out that the equation does not include a term for the effects of gravity,

$$- \frac{V\rho g dh}{dP},$$

where h = height in gravitational field. The imposition of hydrostatic equilibrium reduces the equation to the isentropic expression. On the other hand, the Joule–Thomson equation is applicable to extrusion of magma through a narrow constriction. [Attention was also called to possible geophysical applications of the Joule–Thomson effect by Adams as early as 1922 and by Bowen in 1928 (p. 183).]

PARAMETER PERTURBATION

THERMAL CONDUCTIVITY PERTURBATION

Several processes of magma generation depend on unusual perturbations in one or more properties. Two examples involving either thermal conductivity or density are presented.

Lubimova (1958) noted the general decrease in thermal conductivity with temperature (Figure 4-8) and suggested that with further increase in temperature radiative processes should cause an increase in thermal conductivity. The thermal conductivity was believed to have a minimum of 0.003 cal/cm sec deg at a depth of 50–100 km. Such low thermal conductivity of the upper mantle reduces the rate at which heat escapes from the interior of the earth. McBirney (1963, p. 6327) realized that the temperature would rise in that region and the "heat accumulation" might be sufficient to bring about partial melting. His

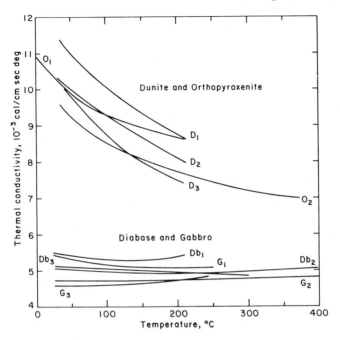

FIGURE 4-8 Change of thermal conductivity of ultrabasic and basic rocks with temperature. From Birch and Clark (1940, p. 549 and 550, Figures 4 and 5). D_1, D_2, and D_3 are dunites; O_1 and O_2 are orthopyroxenites; Db_1 Db_2, and Db_3 are diabases; and G_1, G_2, and G_3 are gabbros. (With permission of the *American Journal of Science*.)

interpretation of Lubimova's statements was supported by the experimental determinations of McQuarrie (1954) and Powell (1954). Confirmation of a minimum in the effective thermal conductivity between 1050°C and 1300°C (Figure 4-9) was obtained by Kawada (1966) for a lherzolite and a dunite and by Murase and McBirney (1970, 1973) for two basalts. (It would be desirable to determine experimentally whether the minimum persists at high pressures.)

McBirney's melting process is probably initiated by high heat flow resulting from a small increase in thermal conductivity produced by an increase in stress (McBirney, 1963, p. 6325), as shown in Figure 4-10. The stressed zone channels the heat flow into a lower temperature zone in the range 1050–1300°C, where the conductivity is low. When the "heat accumulation" is sufficient, the temperature of the ambient rocks, already near the beginning of melting, is raised and melting begins.

There is a "runaway" character to McBirney's minimum thermal

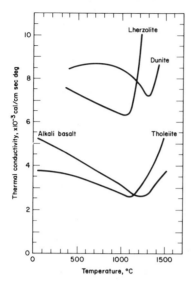

FIGURE 4-9 Thermal conductivity change in lherzolite and dunite (Kawada, 1966, p. 1084, Figures 5 and 6) and two basalts (Murase and McBirney, 1973, p. 3586, Figure 28) with temperature. Note the thermal conductivity minimum between 1050° and 1300°C in both ultrabasic and basic rocks. (With permission of the Earthquake Research Institute, Tokyo University, and the Geological Society of America.)

conductivity melting process (McBirney, 1963, p. 6325). As the lower-temperature rocks are heated, the thermal conductivity decreases and the temperature rises. With increase in temperature, the thermal conductivity decreases further. The temperature rises until the minimum thermal conductivity is reached. In this way, a thermal trap is generated for magma production.

DENSITY PERTURBATION

A contrast in density in a gravity field may also bring about a chain of events leading to melting. Ramberg (1972) considered the behavior of a single buoyant layer a few kilometers in thickness with a density contrast, and suitable viscosity contrast, at a depth of 1,000 km. The process is initiated by a perturbation in the form of a half wave in the layer, extending it to shallower depths (Figure 4-11). The material at the crest of the wave is accelerated upward because of the density contrast with its new surroundings. The detachment characteristics of the mass, a diapir, have been demonstrated by Grout (1945) and Whitehead and Luther (1975) by tank experiments in which inverted drops with long tails were observed. Many of Ramberg's (1972) centrifuge experiments with greater body forces, as well as some of those of Whitehead and Luther, exhibit a massive connection of the diapir to the buoyant layer. According to Ramberg, a thermally isolated mass

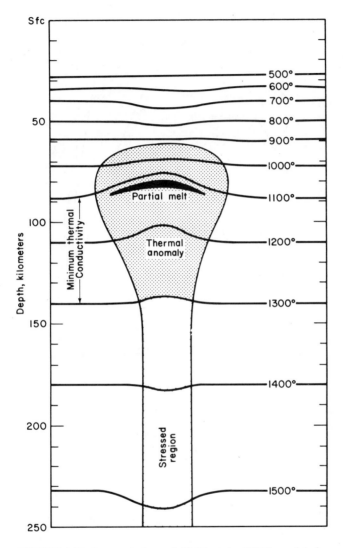

FIGURE 4-10 Interpretation of McBirney's (1963) model for magma generation resulting from the thermal conductivity minimum exhibited in Figure 4-9. A small increase in thermal conductivity, resulting from abnormal stresses, channels the heat flow from depth.

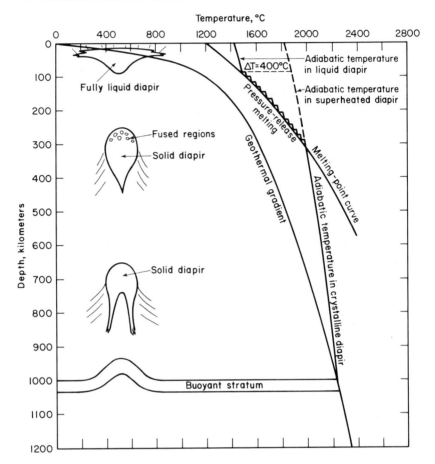

FIGURE 4-11   Magma generation model of Ramberg (1972, p. 58, Figure 10) based on perturbation in buoyant stratum, diapiric rise of a solid mass, and pressure-release melting. (With permission of the Elsevier Publishing Company.)

rises as a solid until the pressure is sufficiently reduced for melting to begin at a depth of about 300 km. Ramberg considered the adiabatic decompression adequate to achieve complete melting when the mass reaches about 80 km. At the maximum rate of rise predicted by Ramberg, it would take about 400,000 years for the mass to travel the 1,000 km to the surface if the viscosity is 10^{22} poises, the density contrast is 0.3 g/cm³, and the initial layer thickness is about 10 km.

Phase changes, including the appearance of liquid, would no doubt accelerate the process at shallower levels as the density contrast increases, whereas cooling of the mass and higher viscosity of the ambient rocks would retard the process.

The diapiric model rests heavily on the dynamic instability of a buoyant layer. If the diapirs were to originate at shallower depths than those considered by Ramberg, the low-velocity layer is an obvious nominee. Contrarily, Press (1970) found the density to be high in that region, and the instability indicated is for the sinking, not the rise, of a mass. In addition, the reader is reminded of the rarity of gabbroic batholiths. A detailed analysis of the heat losses from a diapir would be valuable in determining its potential as a magma source at various depths en route to the surface (*cf.* Cawthorn, 1975, p. 116).

MECHANICAL ENERGY CONVERSION TO HEAT

Another group of mechanisms for generating magma depends on the conversion of mechanical energy into heat. These mechanisms have been called shear melting, frictional melting, tidal deformation melting, viscous heating, and stress-difference melting. The idea stems from Nutting (1929), who calculated the force required to move one surface of rock over another in the absence of deformation and grinding. The work done per unit area is presumed to be converted entirely into heat: $dq = k F dx$, where dq = heat generated, k = coefficient of friction, F = force, and dx = displacement. None of the energy is released as seismic waves. For example, if the force (F) pressing two rock layers together is 1 kbar, the displacement (dx) = 1 cm, and the coefficient of friction (k) = 0.20, then the heat generated (dq) is 4.8 cal/cm^2. This amount of heat is sufficient to raise the temperature of a layer 1 cm thick, having a specific heat of 0.30 cal/g/deg and a density of 2.5 g/cm^3, about 6.4°C.

Experimental evidence has been obtained indicating that the temperature produced along sliding surfaces will approach the melting point of minerals. Bowden *et al.* (1947) observed local transient temperatures up to 1000°C at moderate loads and speeds, using glass, quartz, and other transparent solids through which the transmitted infrared radiation could be measured. Most of the corroborative experiments were carried out on metals that are amenable to a greater variety of techniques (e.g., Bowden and Thomas, 1954). Whereas the true contact area is critical to heat generation, Reitan (1968) also emphasized the need for strain rates greater than 10^{-14}/sec.

THRUST FAULTING

DeLury (1944) applied Nutting's idea to a 1-km overthrust of an outer layer of the earth 60 km thick, which he believed would develop enough heat to warm and melt 200 m at the base of the overthrust. He noted that the Nutting formula does not apply after melting starts and the friction is diminished, and he further emphasized that the speed of deformation and insulating qualities of the rock are important.

SUBDUCTING PLATE

With the development of plate tectonics, the DeLury application was investigated in great detail. For example, Toksöz et al. (1971) calculated the temperature as a function of time in a subducting plate, taking into account phase changes, radioactive heating, adiabatic compression, and shear–strain contributions (Figure 4-12). They believed that with a velocity of transport of 8 cm/yr and a high value of shear–strain heating, melting would begin to take place along the upper surface of a slab 100 km thick at a depth as shallow as 180 km. These magmas were

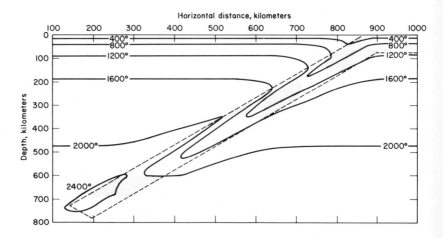

FIGURE 4-12 Magma generation model of Toksöz et al. (1971, p. 1125, Figure 7) based on a high value of shear–strain heating along the upper surface of a 100-km-thick slab subducted into the mantle. The temperature distribution is calculated at time = 13 m.y. with a spreading rate of 8 cm/yr. The stippled areas indicate zones of phase changes. The zone of shear–strain heating is alleged to occur along a narrow zone (~10 km thick) at the top of the slab. Beginning at a depth of about 180 km, melting initiates at the top surface of the slab. In the presence of water, the zone of melting is ±100 km from the top surface of the slab. (With permission. Copyrighted by the American Geophysical Union.)

probably derived from a significant portion of the upper part of the descending slab, consisting of metamorphosed sediments and oceanic basalt; that is, they are probably not basaltic. Temperatures more appropriate for the production of basaltic magma are not achieved until about 200 km, a depth greater than that necessary for magma production in nearby mantle at higher temperature!

In an unsuccessful attempt to explain the high heat flow then believed to occur behind the trench in the sea floor above the subducting plate, McKenzie and Sclater (1968, p. 3177) estimated the heat flow due to stress heating. Using the same velocity of subduction (8 cm/yr) used by Toksöz *et al.*, they obtained a value of 0.6 μcal/cm^2/sec, which is 1/25 that estimated for the heat-flow contribution due to shear by Toksöz *et al.* (1971, p. 1117). In view of the uncertainties, these values are in reasonable agreement.

The principal attraction of frictional melting is that it appears to fulfill the need for magma production directly below a line of active volcanoes paralleling a trench. The lower temperatures relative to adjacent mantle are believed to be adequate if water is available—a point considered below (pp. 78–86). Factors that raise doubts as to the effectiveness of frictional heating as the main cause of magma production associated with the Benioff zone include the presence of anhydrous basalts along the active volcanic line, the increase in age behind the trench, the primitive character of the strontium isotopes of some of the extrusions, the range of composition of layers from quartz- to nepheline-normative, the reduction of friction with initiation of melting, the loss of energy due to seismic events, and the increased lateral heat conduction with depth. The subducting slab is a heat sink and not a likely place to generate magmas.

PROPAGATING CRACK

The shear-melting hypothesis was proposed by Griggs (1954) to account for deep earthquakes, the release of stored elastic energy. The idea was expanded by Griggs and Handin (1960, p. 360–364) and Griggs and Baker (1969), who equated Starr's (1928) calculation for energy released from a propagating crack to the energy required to melt a mass. The aspect ratio of a circular flaw having the appropriate energy for melting is

$$\frac{r}{d} = \frac{4}{3} \frac{\mu\rho\Delta H}{\tau^2} ,$$

where r = radius of flaw, d = thickness of mass around flaw melted, μ = shear modulus, ΔH = enthalpy of melting and τ = shear stress. For $\mu = 3 \times 10^{11}$ dyne/cm^2, $\rho = 3.3$ g/cm^3, $\Delta H = 100$ cal/g $= 4.2 \times 10^9$ dyne cm/g, and $\tau = 100$ bars $= 10^8$ dyne/cm^2, the aspect ratio, r/d, is 5.5×10^5. In order to initiate shear melting the flaw must obviously have an exceptionally high aspect ratio. Griggs and Handin (1960) thought that (a) weak tabular minerals such as mica, (b) a lenticular fluid inclusion, or (c) an instability in structural elements generated by non-Newtonian viscous flow—i.e., creep (see Chapter 9)—would serve as a focal region for initiating shear melting.

REGENERATIVE FEEDBACK

Melting as the result of shear requires that the layer of rock have low heat conductivity so that the heat will not be dissipated. In addition, the heat must be concentrated to provide the large enthalpy of melting. The concentration is presumed to take place because of the nonlinear relationship between viscosity and temperature. In mantle materials the viscosity, η, is experimentally dependent on temperature (Stocker and Ashby, 1973; Kirby and Raleigh, 1973) according to the relationship

$$\eta = \eta_0 e^{-a(T - T_0)},$$

where a varies inversely as the activation energy in rough approximation to the Arrhenius equation (Anderson and Perkins, 1975, p. 118). Because of this relationship, a rise in temperature lowers the viscosity with a resultant increase in strain rate. This sequence, in turn, results in a larger temperature increase, which again increases the strain rate, and so on until the energy source is exhausted. The "runaway" character of this process was described by Griggs and Handin (1960, p. 361) and labeled by Orowan (1960, p. 341) an "avalanchelike" process. A rigorous analysis of several cases was carried out by Shaw (1969, 1973), who used the Gruntfest (1963) theory describing "thermal feedback." The adiabatic case, illustrating a sudden temperature rise with time, is given in Figure 4-13. With a constant stress of 100 bars acting on a solid at its beginning-of-melting temperature and having an initial viscosity of $\eta_0 = 10^{21}$ poises, a runaway condition is reached in about 3.5 m.y.

The regenerative nature of the process, even in large bodies, was emphasized by Anderson and Perkins (1975), who noted its similarity to the exothermic conditions achieved in explosives. The size of the

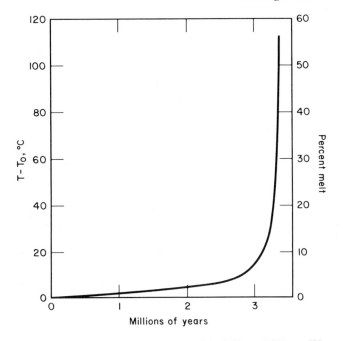

FIGURE 4-13 Magma generation model of Shaw (1969, p. 528, Figure 8) based on accelerated heating resulting from a constant stress on crystalline rock initially at its minimum melting point. Shear heating produces a rise in temperature, which lowers the viscosity, thereby increasing the strain rate. The sequence of events results in a rapid rise in temperature. Model calculated for a 200°C melt range, an initial viscosity of 10^{21} poises, a linear relationship of viscosity with crystal–liquid fraction, and a constant stress of 100 bars. (With permission of the Oxford University Press.)

source region or schlieren zone through which the shear is converted into the enthalpy of melting for a Hawaiian volcano was estimated by Shaw (1973, p. 1516) to be about 5×10^5 km³ ($50 \times 75 \times 140$ km). Such a mechanism for melting no doubt requires mantle deformation closely related to plate motions.

TIDAL DISSIPATION

Shaw (1970) proposed that the tidal energy dissipated in the solid earth, $\sim 10^{19}$ erg/sec, could also be converted to thermal energy. On the basis of his enthalpy of melting of 100 cal/g and his estimate of tidal power, about 30 km³/yr of rock at the melting temperature could be converted to magma by that form of shear melting. That amount of magma is about three to five times the present annual production of lava. In view of Swanson's (1972; personal communication, 1975) estimate that, for present-day Hawaiian eruptions, only roughly 50 percent of the magma reaches the surface, present-day magma could be produced by the complete conversion of tidal energy.

COMPOSITIONAL CHANGE

A major group of processes for melting rocks depends on change of bulk composition. Consider a rock at a temperature somewhat below its beginning of melting. If the composition of the rock is changed, by the addition or subtraction of material, to a new composition whose solidus lies below the same temperature, then melting will ensue. The change of composition can be achieved by metasomatic processes such as diffusion and volatile fluxing.

DIFFUSION

Bell and Mao (1972) used diffusion in a pressure gradient to obtain a more favorable composition for melting. They have suggested differential stress coupled with differential diffusion rates as critical conditions for magma generation. They chose a binary system, R–S, with four compounds related by one peritectic and two eutectic points to illustrate the process (Figure 4-14A). The chemical potential μ and the concentration X for both components are shown in Figure 4-14B in a pressure gradient $P_0 - P_1$. Before diffusion, the composition represented by m in Figure 4-14A would consist of the compounds II and III. If the diffusion of component R greatly exceeded that of component S in response to the pressure gradient, monomineralic layers would develop.

The chemical potentials and concentration after diffusion are given in Figure 4-14C. Because the temperature of point m is above the temperature of the eutectics c and d at P_0, melting would begin at the interface of layers III and IV as well as I and II. Bell and Mao (1972) emphasized that the two derivative liquids generated from a single parent cannot be related to the parental material by fractionation trends and that the two contrasting magmas do not have a simple relationship to each other. The bimodal character of igneous rocks (Daly, 1925b; Chayes, 1963) could be accounted for by the process suggested by Bell and Mao (1972).

Diffusion in Hofmann's (1974) view is limited even when a melt is

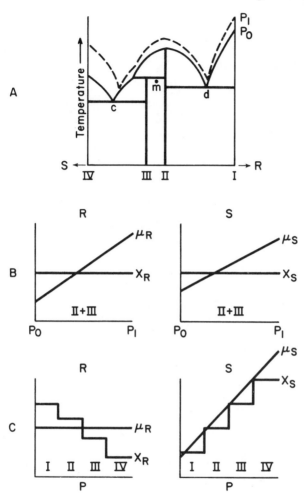

FIGURE 4-14 *A.* Schematic phase diagram for a portion of the system
R–S with four compounds I, II, III, and IV. Two eutectics, having the
compositions *c* and *d*, and a peritectic are exhibited. The composition *m*
consists of compounds II and III. P_0 = liquidus at initial pressure; P_1 =
liquidus at higher pressure.

B. The chemical potential, μ, and concentration, X, of *R* and *S* components as a function of pressure *before* diffusion.

C. The chemical potential, μ, and concentration, X, of *R* and *S* *after*
diffusion, assuming the diffusion of *R* greatly exceeds that of *S*. Four
monomineralic zones are produced. Melting begins at the interface of zones
I and II and at the interface of zones III and IV according to Bell and Mao
(1972, p. 418, Figure 28). (With permission of the Carnegie Institution of
Washington.)

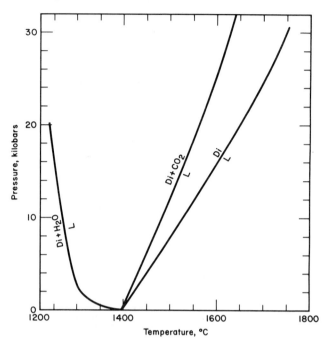

FIGURE 4-15 Melting curves of diopside (Di) without volatiles (Yoder, 1952; Boyd and England, 1963), with H_2O (Yoder, 1965, p. 87, Figure 12; Eggler, 1973, p. 458, Figure 17), and with CO_2 (Eggler, 1973, p. 458, Figure 17). (With permission of the Carnegie Institution of Washington.)

present to "the scale of a few hundred meters," and therefore the diffusion-melting process would require an extensive zone of shearing to produce large volumes of magma. The more rapid diffusion of volatiles, however, is another matter.

VOLATILE FLUXING

Volatiles have a considerable influence on melting behavior. Both H_2O and CO_2, the principal volatiles in the earth, reduce melting temperatures well below that for melting in the absence of volatiles. A representative measure of the relative influence of H_2O and CO_2 on the melting temperatures can be seen in Figure 4-15 (see also Figure 10-12) for the major rock-forming mineral diopside. One may easily estimate the effect of adding either H_2O or CO_2 to a rock near the beginning-of-melting temperature.

It is not necessary that the volatile be present in amounts in excess of that required to saturate the liquid. Although only one system, albite–water, has been studied in sufficient detail to document the principles, that system is thought to be representative of a wide variety of rock-forming minerals in the presence of volatiles. The effect of reducing the water pressure below the total pressure is displayed in Figure 4-16. An important observation is that the beginning of melting

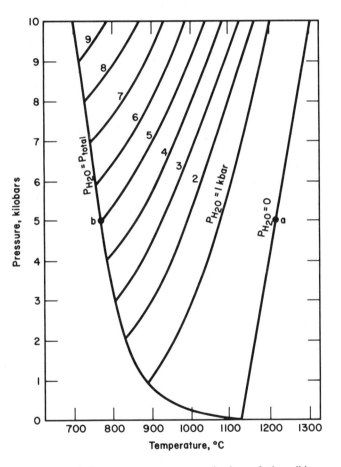

FIGURE 4-16 Pressure-temperature projection of the albite–H_2O system with the liquidus indicated for various water pressures (after Burnham and Davis, 1974, p. 936, Figure 19). The temperatures *a* and *b* are displayed in an isobaric section in Figure 4-17. (With permission of the *American Journal of Science*.)

remains the same whether or not H_2O is available in excess of that required to saturate the initial liquid. The liquidus, on the other hand, is lowered below the anhydrous liquidus as the water pressure is increased. The temperature–composition section for a total pressure of 5 kbar is given in Figure 4-17. The points *a* and *b* are identified in Figure 4-16. If the amount of water present is 0.5 percent, the beginning of melting is the temperature of *b* and the water content of the liquid is about 10 percent, but only 5 percent liquid is produced. The amount of

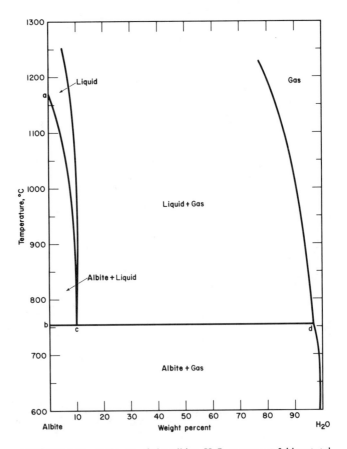

FIGURE 4-17   Estimate of the albite–H_2O system at 5 kbar total pressure. The temperature of *a* is from Birch and LeComte (1960); *b* and *c* are from Yoder *et al.* (1957, p. 208, Figure 38) with correction by Morse (1970, p. 234, Figure 2; Errata). The composition of *d* is from Davis (1972, p. 31, Figure 1A). (With permission of the Carnegie Institution of Washington.)

water present, in effect, determines the proportion of liquid to crystals. If 2.5 percent H_2O is present, then 25 percent of the albite + H_2O mixture will initially melt at the temperature *b*. As the temperature increases above *b*, the liquid contains decreasing amounts of H_2O.

The melting of a hydrous phase introduces a variety of complexities. A temperature-composition section is shown in Figure 4-18 for the anhydrous constituents of the phlogopite (= forsterite + kalsilite + leucite) and water system (Yoder and Kushiro, 1969). The compositional region the writer believes to be most pertinent to the melting of rocks in the mantle, where the H_2O content is presumed to be very low, is unfortunately not well determined. (Additional hydrous systems in the absence of a free gas phase should be investigated, but the rates of reaction are slow and equilibrium is difficult to establish.) The diagram implies unique melting relations different from those where the H_2O content exceeds that of the hydrous crystalline phase. If the higher beginning-of-melting temperature is a general effect in assemblages consisting of anhydrous and hydrous phases only (without a free gas phase), then present melting data on minerals in the presence of excess H_2O may not be pertinent to melting in the mantle. For this reason, melting in the mantle may not be as pervasive as the relations in Figure 4-20 suggest. As will be seen later, the presence of phlogopite in the mantle is believed to be necessary because of its potassium and water content. None of the *major* minerals in garnet peridotite is known to carry sufficient potassium to yield basaltic compositions on melting (see Kushiro, 1973b, p. 295, Table 84), and phlogopite has the largest range of stability known for a hydrous mineral. On the other hand, there are limitations to the amount of modal phlogopite because of the heat production constraints on the potassium content of the mantle.

When both CO_2 and H_2O are present, both the liquidus and solidus are dependent on the proportions of the volatiles. In Figure 4-19 is shown the influence of various proportions of CO_2 and H_2O on the solidus and liquidus of the diopside system. The effect was also observed in a study of a natural garnet peridotite by Mysen (1973, p. 473).

Some investigators believe that there is always a gas phase present in the mantle because of the small amount of CO_2 required for saturation of magma. For example, only 4.8 ± 1.0 percent CO_2 (Eggler *et al.*, 1974, p. 227) is required to saturate a diopside melt at 30 kbar and 1625°C, whereas 21.5 ± 1.0 percent H_2O (Hodges, 1974, p. 253) is required at the same pressure and 1265°C. Most workers, however, have the opinion that there is not a free gas phase and the volatiles are stored in various hydrous and carbonate minerals. Most of the hydrous

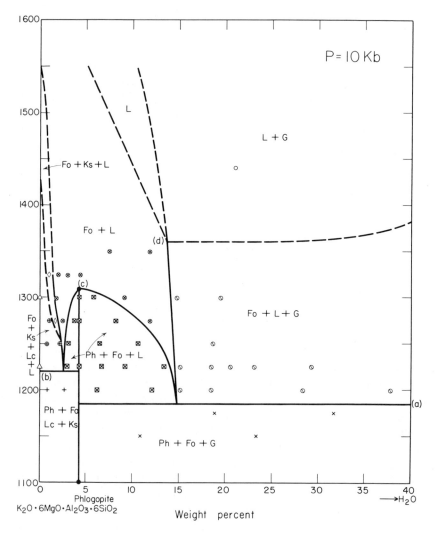

FIGURE 4-18   Temperature–composition projection at 10 kbar for the silicate-rich portion of the $K_2O \cdot 6MgO \cdot Al_2O_3 \cdot 6SiO_2 - H_2O$ join, illustrating the complex melting relations of the hydrous phase phlogopite in the absence of a free gas phase. Ph = phlogopite; Fo = forsterite; Lc = leucite; Ks = kalsilite; L = liquid; G = gas. Temperature a is the beginning of melting in the presence of excess H_2O; b, the beginning of melting in the absence of a gas phase; c, the maximum thermal stability of phlogopite; and d, the minimum liquidus in the absence of a gas phase. The melting relations for the assemblage $Ph + Fo + Lc + Ks +$ are closely analogous to the assumed relations in the mantle. From Yoder and Kushiro (1969, p. 576, Figure 4). (With permission of the *American Journal of Science*.)

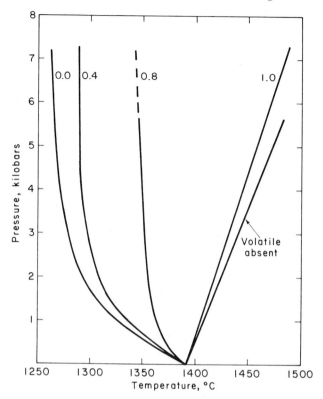

FIGURE 4-19   Melting of diopside in the presence of CO_2 and H_2O and in the absence of volatiles. The solidus is raised as CO_2/$(CO_2 + H_2O)$ in the vapor, indicated on each curve, increases according to Rosenhauer and Eggler (1975, p. 475, Figure 47). (With permission of the Carnegie Institution of Washington.)

minerals break down at temperatures below melting, and the volatiles, if retained, are eventually consumed in the melt; only a few hydrous minerals are stable into the melt region. Phlogopite, for example, is not an uncommon constituent in ultrabasic rocks. In addition, there is the possibility that garnets and pyroxenes store water as hydrogarnet and hydropyroxene by way of substitution of $4H^+ \rightarrow Si^{4+}$. At pressures less than about 30 kbar, various amphiboles are stable. Polymorphs of calcite and dolomite are stable to very high pressures. The melting curve of calcite (Yoder, 1975b, p. 892, Figure 6; Irving and Wyllie, 1975, p. 38, Figure 2) is about 100° below that of diopside. The paucity of diopside relative to enstatite in the mantle may be attributed to a small degree to the reaction

diopside + 2 forsterite + 2 CO_2 = 4 enstatite + dolomite,

experimentally determined by Eggler (1975, p. 472, Figure 46). The concept that the earth is continuously degassing was supported in a detailed analysis by Rubey (1951). If the major oceans were formed by degassing primarily during the Precambrian, then there should be vast regions in the mantle depleted in volatiles. The observation of great volumes of relatively anhydrous basalt on the present ocean floor supports that view.

Over the years, the water content of the mantle has been estimated at various values ranging from 13 wt% (the H_2O content of serpentine) to zero. In general, opinion about water in the mantle has swayed from excess H_2O to H_2O deficiency (free vapor absent and water insufficient to saturate a magma), but with the increased role assigned to CO_2 the idea of a free vapor phase has been revived. Others hold to the concept of a relatively anhydrous mantle with volatiles, especially H_2O, being added to the magma in the crust while stored in an auxiliary magma chamber. The hydration of a magma by diffusion, however, is probably limited to within a few tens of meters of the boundary (Shaw, 1974, p. 156). The oxygen and hydrogen isotopes present strong evidence that H_2O circulating at the periphery of shallow chambers is absorbed for the most part *after* crystallization of the magma (Taylor, 1974, p. 856). Although the criteria for identifying primordial H_2O are not well defined, anomalously high $^3He/^4He$ in seawater (Clarke et al., 1969), in pillow basalt glass rims (Lupton and Craig, 1975), and in volcanic gases (Tolstikhin, 1975) has led to the view that primordial ^3He is being released from the mantle. The implication is that there may be a continuous flux of primordial H_2O from the undepleted mantle. Although most of the volatiles associated with basalt and andesite are considered to be of crustal origin, those volatiles in kimberlite appear to offer an opportunity to test a mantle origin.

The degassing of the earth, if that is indeed the source of the oceans, must have resulted in extensive metasomatism of the upper mantle. One cannot help being impressed by the data of Nakamura and Kushiro (1974, p. 258) on the composition of gas in equilibrium with forsterite and enstatite. The SiO_2 content of the *gas* is of the order of 20 percent (wt) at 15 kbar and 1150–1350°C! Although the presence of CO_2 may reduce the solubility of SiO_2, such an ascending gas would greatly enrich the upper horizons in enstatite relative to olivine. The other constituents in the gas would produce an array of other changes, perhaps including the deposition of secondary phlogopite. It is no wonder that most of the nodules retrieved from basalts and especially from kimberlites appeared to have been soaked in a corrosive juice and metasomatized. As will be seen below, volatiles may also play a major role in determining the kind of magma that reaches the surface of the earth.

The principles outlined above can be applied to the melting of peridotite. Kushiro *et al.* (1968a) and Mysen and Boettcher (1975) studied the melting of natural spinel lherzolite in the presence of H_2O with and without CO_2 (Figure 4-20). In the presence of an excess of H_2O, the solidus is curve F and a large partial melt region would exist below about 50 km under the oceans and below about 80 km under the Precambrian shields, assuming the geotherms of Clark and Ringwood (1964) are applicable. With increasing amounts of CO_2 (curves E, D, and C), the solidus is raised and melting begins at successively greater

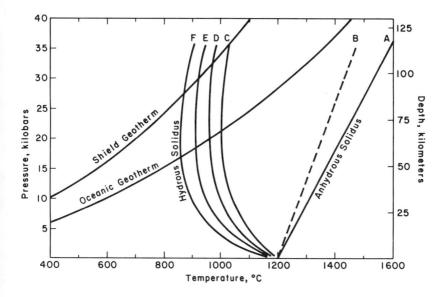

FIGURE 4-20   Melting relations of a natural spinel lherzolite. Curve A is the solidus under anhydrous conditions (Kushiro *et al.*, 1968a, p. 6026, Figure 1); curve B, the estimated solidus with CO_2; curves C, D, and E, respectively, the solidus with $H_2O/(CO_2 + H_2O)$ = 0.25, 0.50, 0.75; curve F, the solidus with only excess H_2O (Mysen and Boettcher, 1975, p. 542, Figure 6B). Geotherms from Clark and Ringwood (1964, p. 53, Figure 2). (With permission. Copyrighted by the American Geophysical Union and Oxford University Press.)

depths under both oceanic and shield regions. The results with H_2O have been verified for the most part by Kushiro (1970), Green (1973), and Millhollen et al. (1974).

The excessive amount of melting in the presence of an excess of water was an embarrassment to the early experimenters applying their results on minerals to magma production in the mantle. Some of the conclusions drawn were that (a) water was not present in excess in the mantle—it was not a free phase; (b) the assumed geothermal gradient may be too high in temperature for the upper mantle; (c) peridotite may not be the parental material in the mantle; (d) the mantle was depleted of volatiles during the early Precambrian to form oceans; and (e) other volatiles may reduce the effectiveness of water. The present preferred view is that the mantle is not uniformly or entirely depleted of volatiles, and whereas water is the principal volatile, other volatiles such as CO_2 reduce the activity of water. Given the present range of estimated geotherms, which do not intersect the anhydrous solidus of peridotite, it seems clear that some small amount of volatiles must be present if the low-velocity zone now observed is caused by partial melting. The amount and proportion of volatiles will determine the amount of melt now existent.

OTHER MECHANISMS

One of the more important mechanisms for the production of melt is radioactive heating, as will be discussed in the following chapter. The role of exothermic chemical reactions was considered important by Daly (1911), Day and Shepherd (1913), and Jaggar (1917); the present writer, however, supports the view of Holmes (1927) that such reactions are probably not of fundamental importance until the magma has accumulated and is rising to the surface.

PHENOMENOLOGICAL VIEW OF MELTING

Before leaving the various mechanisms for producing a melt, it is appropriate to mention the potential opportunities for investigating the physics of melting of the crystals that make up the rocks. The change from a crystalline state to a liquid state is considered by some investigators to result from the breaking of the weakest bonds in a crystal. As the melting point of a crystal is approached, increasing amounts of energy are absorbed (see Figure 5-1) by the crystal through vibrational stretching of the bonds (Lindemann, 1910). When the stretching exceeds a critical value, the bond breaks and melting ensues. The resulting masses of substructural units dispersed in a low-strength medium appear to have exotic compositions.

Positional disordering and other mechanisms for increasing the entropy are favored by some investigators in describing the change from the crystalline to the liquid state. The liquid state is believed to be quasi-crystalline and is characterized by loss of long-range ordering. Such models are amenable to statistical analysis; at best, however, they are somewhat artificial, according to Ubbelohde (1965, p. 201). On the basis of the Tolland and Strens (1972) model for electrical conduction, Strens (personal communication, 1975) described melting as the critical stage where a three-dimensional net of dislocations obtains and the masses of multiple unit cells fail in strength.

The structure of diopside was determined with precision at a series of temperatures up to 1000°C (Cameron *et al.,* 1973; Finger and Ohashi, 1976), and there were no significant distortions of the structures with increasing temperature. The vibrational ellipsoids do not intersect when their changes are projected to the melting "point," and there is no obvious indication that the structure is about to fail because of mechanical instability. Hazen (1975, p. 171) concluded that the melting points of olivines may be determined by structural constraints. He observed that forsterite, hortonolite, and fayalite all have similar cell constants, hence a maximum cell volume, at their respective melting points. Determination of the crystal structure in the region of anomalous increase in enthalpy prior to melting and of the liquid structure just above the melting point is a goal almost within reach of present technology. A satisfactory theory is needed to explain the phenomenon of melting and to account for the enthalpy of melting.

5 Thermal Energy Requirements for Melting

Volcanic heat represents only a very small fraction of the heat radiated by the earth.

. . . the real problem of volcanoes is not so much to find a suitable source of energy as to provide ways and means by which rather insignificant amounts of heat can be focussed on relatively insignificant volumes of the crust.

Verhoogen (1946, p. 747)

ENTHALPY OF MELTING

The thermal energy required to convert crystalline rock to magma, the enthalpy of melting, is often referred to as "heat of melting," "heat of fusion," or "latent heat of fusion" even though heat is defined as the energy transferred as a result of a temperature difference (Zemansky, 1937, p. 48).* Why is it important to know the amount of energy required to bring about melting? First, the enthalpy may be a guide to the distribu-

---

* Heat is not a quantity describing a state of a system, nor is it the stored energy inside a system. In Professor G. Tunell's view (personal communication, 1975), "The amount of heat that would be transmitted in a strictly isothermal expansion or compression" can be calculated even though heat does not flow from one body to another in the absence of a temperature difference. He has considered the enthalpy of melting "a limit value approached when the temperature difference between surroundings and a pure, simple compound in equilibrium with its liquid is very small." There is some basis, therefore, for using the common expressions "heat of melting," "heat of fusion," or "latent heat of fusion" in describing the thermal energy required for melting a rock.

88

tion of heat-producing sources. Second, it is a constraint on the mechanism of magma formation. Third, enthalpy is an important factor in determining the amount of time required to produce a melt, and the rate of magma production may relate to the periodicity of eruption (see Chapter 10). The amount of energy required to bring a rock to the melting temperature may be considerable, depending on the initial temperature, and some estimates for simple systems are given below. During heating to the melting temperature, a rock may undergo several endothermic metamorphic reactions. The total energy consumed in these reactions may far exceed the energy required for melting. For most purposes the ambient temperature is presumed by petrologists to be already below but near the beginning of melting, so the principal concern is the source of energy for the melting process itself.

MEASUREMENT

From a practical standpoint the observation of the enthalpy of melting in a differential thermal apparatus is made over a temperature interval, and the energy absorbed appears as an anomalous heat capacity (Figure 5-1). The area under the peak, when properly calibrated, is a measure of the enthalpy of melting. The enthalpy of melting is usually measured more accurately in a calorimeter by observing the difference in energy released by dropping a mass of crystals and a separate mass of glass of equivalent composition, respectively, from temperatures just below and just above the melting point into a relatively cool, thermally insulated bath. The rise in temperature of the bath or the change in proportion of crystals and liquid in a suitable bath mixture is used to evaluate the energy released by each state of the material.

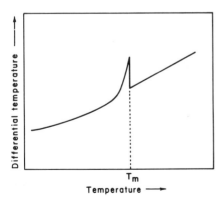

FIGURE 5-1 Schematic representation of the differential thermal analysis of a silicate undergoing congruent melting. The area under the peak, enclosed by extrapolating the baseline, is proportional to the enthalpy of melting. T_m = melting temperature.

Early measures of the enthalpy of melting, ΔH_m, were made, for example, by Barus (1893), who obtained a value of 24 cal/g (1200°C) for a diabase using the cooling-curve method. Vogt (1904, p. 55) considered this value to be too low because the diabase did not crystallize completely on cooling. He preferred a value of 90–100 cal/g, based on the enthalpy of melting of minerals such as diopside, anorthite, akermanite, and fayalite. Bowen (1928, p. 314–315) assumed that the value 100 cal/g was roughly that for basalt, presumably relying on the values for diopside and anorthite obtained from calculations of the freezing-point depression as well as measured values for the constituent minerals (e.g., White, 1909, p. 486, last footnote). He later found support for using this value even for peridotites when Bowen and Schairer (1935, p. 207) estimated that the ΔH_m for pure forsterite was exactly 100 cal/g on the basis of calculations of the freezing-point depression in the forsterite–fayalite system. The value 100 cal/g has been used in many present-day calculations of the thermal factors involved in the generation of basaltic magma.

ENTHALPY OF MELTING OF MINERALS AND ROCKS

VALUES AVAILABLE FOR END-MEMBER MINERALS

Although it is not yet possible to deal effectively with iron-bearing natural rocks in high-temperature calorimeters because of oxidation and crystallization problems, estimates of the enthalpy of melting can be made using data on constituent end-member minerals. The values now available are given in Table 5-1 (Yoder, 1975a, p. 515).

MODEL SYSTEM FOR BASALT

The calculated enthalpy of melting for the "eutectic" composition, $Di_{58}An_{42}$ (wt%), of the diopside–anorthite system (Figure 5-2), neglecting solid solution, phase changes, and heat of mixing, is 77.8 cal/g. Using values for the heat content of the end-member phases from Robie and Waldbaum (1968, p. 221 and 226), the heat requirements to melt compositions in the diopside–anorthite system can be displayed in a plot of the change in enthalpy, H_T-H_{298}, versus composition (Figure 5-3). Such a diagram can be used as a model for some aspects of the melting of basalt.

The data in Figure 5-3 illustrate how small ΔH_m is relative to the *total*

TABLE 5-1   Enthalpy of Melting[a]

End-Member Mineral	ΔH_m (kcal/mol)	Formula wt	ΔH_m (cal/g)
Forsterite	29.3 (est)[b]	140.70	208.2
Fayalite	22.0_3	203.78	108.1
Clinoenstatite	14.7	100.38	146.4
Diopside	18.5	216.52	85.4
Pyrope[c]	33.2^d	403.08	82.4
Anorthite[c]	18.7^d	278.14	67.2
High albite	13.5_6	262.15	51.7
High sanidine	14.7	278.24	52.8

[a]From Robie and Waldbaum (1968) unless otherwise noted.
[b]Bradley (1962).
[c]O. J. Kleppa and T. V. Charlu (unpublished data, 1975).
[d]ΔH_m at 700°C.

heat required to bring the assemblage to the temperature of the beginning of melting. Beginning at temperatures assumed for the upper mantle, the amount of heat required for *partial* melting is still relatively small compared with that required to raise the temperature of the rock to the beginning of melting. The figure also indicates how the heat

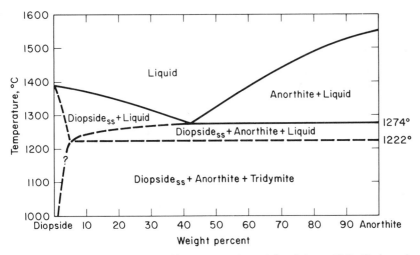

FIGURE 5-2   The diopside–anorthite system at 1 atm (after Osborn, 1942; Clark *et al.*, 1962; Yoder, 1965). The depression of the solidus for diopside-rich compositions is due primarily to solid solution of the Ca-Tschermak's molecule, $CaAl_2Si_2O_6$, in diopside and resultant enrichment of silica in the liquid. [The extensive solid solution (25 wt%) indicated by Hytönen and Schairer (1961, p. 136), on the basis of cell dimensions and optical observations, is considered to be suspect because of metastability.] (With permission of the Carnegie Institution of Washington.)

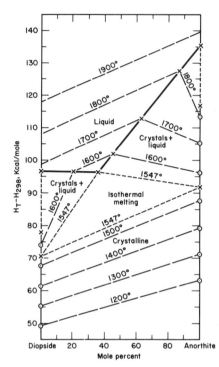

FIGURE 5-3   Plot of $H_T - H_{298}$ versus composition in the diopside–anorthite system, neglecting solid solutions, phase changes, heat of mixing, and pressure effects. Temperature in degrees K. The value of ΔH_m for diopside measured at 973°K was used and not corrected to the melting temperature. Modified after Yoder (1975a, p. 516, Figure 76). (With permission of the Carnegie Institution of Washington.)

requirements change as a function of composition in the isothermal region and through the crystal + liquid regions. Although basalts are eutecticlike at the liquidus because all major phases appear within a small temperature interval (Yoder and Tilley, 1962), these phases, owing to solid solution, continue to crystallize together from liquid over a more extended range of temperature. The temperature range is also a function of the degree of fractionation of crystals and liquid. It is instructive, therefore, to examine melting in a system involving a solid solution series such as forsterite–fayalite. Figure 5-4 was constructed for the most part from the thermodynamic data given by Robie and Waldbaum (1968) and the phase-equilibrium data of Bowen and Schairer (1935). It is seen that, with melting, heat is consumed at a greatly increased rate but without a thermal arrest as for a eutectic system.

IMPROVED MODEL SYSTEM FOR BASALT

The model system for basalt can be improved by choosing the piercing-point composition at 1 atm, $Di_{49.0}An_{43.5}Fo_{7.5}$ (wt%), in the diopside–anorthite–forsterite system, studied by Osborn and Tait (1952). The enthalpy of melting can be calculated from the end-member minerals, assuming that there is no solid solution or heat of mixing. The contribution of the requisite portion of forsterite to the enthalpy of melting is estimated from the calculated value of Bradley (1962) because all techniques tried have failed to produce a glass of forsterite composition for calorimetric measurement. The calculation yields a value of 85.6 cal/g for the piercing-point composition. To obtain an estimate of the heat of mixing, the enthalpy of melting of the piercing-point composition was measured in a high-temperature calorimeter at 700°C by O. J. Kleppa and T. V. Charlu (see Yoder, 1975a). They obtained a value for the enthalpy of melting of 18.25 ± 0.14 kcal/mol based on an average molecular weight of 229.38, or 79.6 cal/g. If the estimate for the enthalpy of melting of forsterite is acceptable and the effects of solid solution can be neglected, the heat of mixing, obtained by difference, is about -6.0 cal/g, a small but not negligible value.

For a model basalt with a more appropriate feldspar composition, the piercing point of the system diopside–forsterite–anorthite$_{50}$–albite$_{50}$ (Yoder and Tilley, 1962, p. 396) may be taken. The composition of this point is Di$_{28.5}$Fo$_{4.5}$An$_{33.5}$Ab$_{33.5}$ (wt%), and its calculated ΔH_m at 1 atm is 73.5 cal/g. The addition of iron to the system would tend to lower the enthalpy of melting but by an unknown amount.

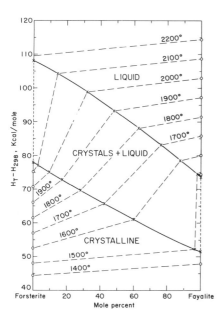

FIGURE 5-4 Plot of H_T–H_{298} versus composition in the forsterite–fayalite system, neglecting heat of mixing, incongruent melting, and pressure effects. Temperature in degrees K. From Yoder (1975a, p. 517, Figure 77). (With permission of the Carnegie Institution of Washington.)

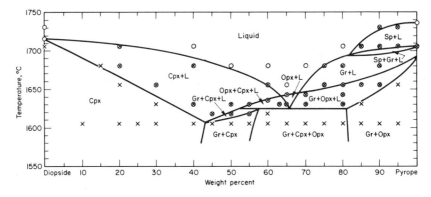

FIGURE 5-5   The diopside–pyrope system at 30 kbar according to O'Hara and Yoder (1967, p. 74, Figure 3). × = crystalline; ⊗ = crystals + liquid; ○ = liquid. *Sp* = spinel. (With permission of Oliver and Boyd.)

MODEL SYSTEM FOR ECLOGITE

In a similar way the enthalpy of melting for the "eutectic" composition of the system diopside–pyrope at 30 kbar (Figure 5-5) can be calculated, neglecting solid solution, heat of mixing, and the presence of orthopyroxene. For this composition, $Di_{34}Py_{66}$ (wt%), the enthalpy of melting calculated for 1 atm is 83.4 cal/g. To obtain the enthalpy of melting at 30 kbar it is necessary to use the relationship

$$\left.\frac{\partial \Delta H}{\partial P}\right|_{T} = \Delta V_{m}(1 - \Delta \alpha T),$$

where ΔV_{m} is the volume change on fusion and $\Delta \alpha$ is the difference between the coefficients of thermal expansion of the solid and liquid phases (see Maaløe, 1973, p. 111). Because of the small value of the coefficients of thermal expansion ($\sim 10^{-4}$/deg), the last term may be neglected. For a ΔV_{m} of about 0.049 cm³,* the change of enthalpy of melting resulting from a pressure increase of 30 kbar is about 35.1 cal/g, a substantial increase! The value for the enthalpy of melting of $Di_{34}Py_{66}$ at 30 kbar is therefore about 118.5 cal/g. The melting of the diopside–

---

* Based on the following densities measured at 1 atm and not corrected for pressure or temperature:

Pyrope: $\rho_{glass} = 3.031$ g/cm³ (B. O. Mysen, unpublished data, 1975)
$\rho_{crystal} = 3.582$ g/cm³ (Skinner, 1956, p. 428)
Diopside: $\rho_{glass} = 2.846$ g/cm³ (Larsen, 1909, p. 271)
$\rho_{crystal} = 3.275$ g/cm³ (Allen and White, 1909, p. 14)

pyrope system may be used as a model for some aspects of the melting of eclogite.

MODEL SYSTEM FOR GARNET PERIDOTITE

Garnet peridotite is herein considered the most likely parent of basaltic magma, so an estimate of its enthalpy of melting is useful. The piercing point or "eutectic"* of the system diopside–forsterite–pyrope at 40 kbar (Figure 5-6) has the composition $Di_{47}Fo_6Py_{47}$ (wt%), and the estimated ΔH_m at 1 atm is 91.4 cal/g. Davis and Schairer (1965, p. 125) estimated that the composition of the invariant point for the diopside–enstatite–forsterite–pyrope system was $Di_{47}En_3Fo_3Py_{47}$ (wt%) at 40 kbar. The ΔH_m at 1 atm for that *reaction* point, estimated from the end-member phases, is 89.5 cal/g. It is not possible to calculate accurately the pressure effect because of lack of knowledge of the volume change of forsterite on melting†; however, a provisional value of the enthalpy of melting at 40 kbar is about 135.4 cal/g. The invariant point is sufficiently close in composition to represent the formation of basaltic liquid. Its norm is $Di_{22}Fo_{20}An_{32}En_{26}$ (wt%). The ΔH_m values, when corrected for the pressure at the site of generation, may be as much as 25 percent higher than the 100 cal/g value used by Bowen (1928) for basalts. On the other hand, the pressure effect may be counterbalanced to some degree by the presence of water, which would be expected to reduce ΔH_m greatly.

---

*The *piercing point* is the intersection of a ternary join, for example, with a univariant curve. The univariant curve is the locus of composition of liquids in equilibrium with the three participating crystalline phases over a range of temperature. If the range of temperature is small, the piercing point is commonly designated as an approximation by the term "eutectic." The univariant curve, or piercing-point curve, joins the eutectic of the multicomponent system. The determination of a piercing point is difficult because equilibrium is not always readily attained at the lowest melting temperatures of a system. Once recognized in a join, the piercing point is an indication that the composition of one or more of the phases cannot be represented wholly by the chosen components.

†A ΔV_m for forsterite can be calculated from the melting-curve slope using the Clausius–Clapeyron equation. If $dT/dP = 4.77°C/kbar$, $T_m = 2163°K$, and $\Delta H_m = 208.2$ cal/g, then $\Delta V_m = 0.019$ cm³/g. The ΔV_m calculated from the melting-curve slope is *usually* considerably less than that based on measured values of the density of glass (or liquid) and crystals (see discussion by Yoder, 1952, p. 369 *ff.*). Using the index of refraction of forsterite glass ($n = 1.634$), estimated by Bowen and Schairer (1935, p. 207, Table VII) by extrapolation from synthetic iron-rich compositions and the specific refractive energies for MgO and SiO₂ of Larsen and Berman (1934), the density of glass is 3.123 g/cm³. The measured density of forsterite is 3.223 g/cm³, and therefore the ΔV_m is approximately 0.010 cm³/g by this method. This disagreement will be resolved when a glass of forsterite is obtained or a direct measurement of ΔV_m is made.

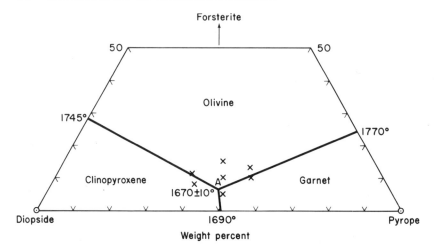

FIGURE 5-6 The liquidus fields of a portion of the forsterite–diopside–pyrope system at 40 kbar. The point *A* is a piercing point of the univariant curve Di–Py–Fo–L leading *from* the invariant reaction point Di–Py–Fo–En–L at a temperature of 1680 ± 10°C, according to Davis and Schairer (1965, p. 124, Figure 35). The two temperatures (1670 ± 10°C and 1680 ± 10°C) are within the error of measurement, and the temperature gradient may be reversed between the piercing point and the invariant point. (With permission of the Carnegie Institution of Washington.)

APPLICATION OF ENTHALPY OF MELTING

To gain an appreciation of the role of the enthalpy of melting, two mechanisms will be examined in detail: (a) consumption of local heat production and (b) adiabatic rise.

LOCAL HEAT PRODUCTION

Consider a layer with a fixed radioactive heat production at 200 km under a continent (Figure 5-7). The heat flow at 200 km is assumed to be 0.60×10^{-6} cal/cm²/sec (Sclater and Francheteau, 1970). [Jordan (1975), on the other hand, preferred a much lower value of 0.21×10^{-6} cal/cm²/sec at 200 km in order that the temperature at that depth be less than 1200°C, which he believed to be appropriate for the partial melting of a peridotite containing 0.1 percent H_2O.] Radioactive heat production changes with time; it is therefore necessary to keep changes of heat flow in mind when considering magma generation during various geologic periods. Lee (1967) illustrated (Figure 5-8) the heat production

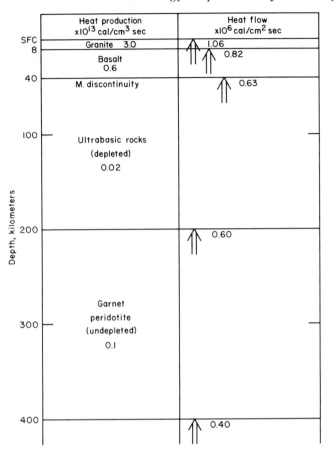

FIGURE 5-7 Heat production and heat flow for a model of a continental shield assuming a constant heat flux at a depth of 400 km. After Holmes (1945, p. 481, Figure 254) and Sclater and Francheteau (1970, p. 525, Figure 12). *SFC* = surface. (With permission. Copyrighted by the Ronald Press Company.)

from important isotopes as a function of time.* The major contributors are uranium, potassium, and thorium, and their average abundances in relevant rocks are given in Table 5-2.

There is no obvious correlation between magma generation and the

---

*The diagram of Lee (1967) was reproduced by Lubimova (1969, p. 65, Figure 1) but contains numerous original drafting errors. Dr. Lee (personal communication, 1976) reconfirmed his calculations, and the curves drafted correctly are reproduced in Figure 5-8 with his kind permission.

TABLE 5-2 Average Abundances of Heat-Producing Elements and Heat Production from Radioactivity in Basic and Ultrabasic Rocks and Model Composition[a]

Rock Group	Reference	Average Abundances of the Radioactive Elements (ppm)			Density	Total Heat Production from Radioactivity (cal/sec/cm³)[c]
		U	Th	^{40}K[b]		
Basaltic rocks	Holmes (1965)	0.7	3.0	1.1	2.91	1.20×10^{-13}
Oceanic tholeiites	Engel and Engel (1964); Tatsumoto et al. (1965)	0.16	0.15	0.14	3.00	0.17×10^{-13}
Ultrabasic rocks	Holmes (1965)	0.013	0.05	0.001	3.33	0.02×10^{-13}
Lherzolites	Wakita et al. (1967)	0.019	0.05	0.007	3.15	0.03×10^{-13}
Pyrolite II	Ringwood (1958)	0.059	0.25	0.09	2.82	0.10×10^{-13}

[a]In part from Sclater and Francheteau (1970, p. 524, Table 5).

[b]^{40}K = 0.0119 percent total K (Nier, 1950).

[c]Present heat production in cal/g/sec (MacDonald, 1959, p. 1969): U: 2.25×10^{-8}; ^{232}Th: 0.63×10^{-8}; ^{40}K: 0.70×10^{-8}.

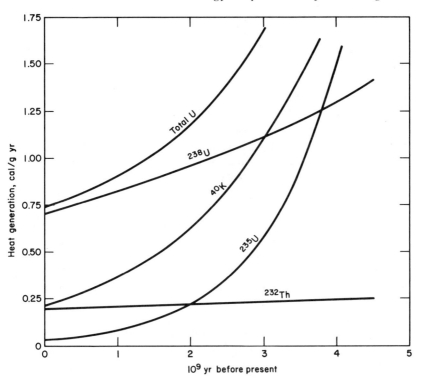

FIGURE 5-8   Heat generation by important radioactive isotopes with geological time (after Lee, 1967; personal communication, 1976).

abundance of radioactive elements. Magmas are not abnormally radioactive, nor are regions high in radioactivity particularly more active volcanically. As indicated in Table 5-2, samples of potential mantle materials have a very low content of radioactive elements. Of course, it is not known to what extent these samples have already been depleted of these elements.

A cubic centimeter of undepleted garnet peridotite just *under* the depleted zone, already divested in part of potential liquids of basaltic composition, generates about 10^{-14} cal/cm³/sec. If the entire heat production is retained, it would take 82 m.y. to supply enough energy to melt just 5 percent of the rock at 200 km ($P = 64$ kbar) assuming that it is already at the beginning-of-melting temperature, the enthalpy of melting is 164 cal/g, and the density is 3.2 g/cm³. Obviously, it would take considerably longer if heat losses were appreciable or the degree of melting were greater. The time required for melting would also be longer in the depleted zone and shorter in the lower crust provided the temperature were raised sufficiently for melting to begin.

Retention of local heat production would influence the heat flow. In order to maintain the heat flow at the surface, the rocks at the site of magma generation would require either (a) a higher than normal radioactive element content or (b) a perturbation in the heat flow from rocks below or adjoining. The higher radioactive element content of basalt relative to its alleged source rocks is presumably the result of the preference for those elements to partition into the basaltic melt (12:1, based on Table 5-2). The experimental data of Shimizu and Kushiro (1973, p. 271) suggest an enrichment in the liquid relative to the starting material for several elements, including potassium, of 5:1 for a ~20 percent partial melt, tholeiitic in major element composition, of garnet lherzolite at 1450°C and 15 kbar. The implication is that basalt is the product of a small amount of partial melting.

McBirney (1967) considered how an initial thin layer of melt would increase in thickness as additional heat was absorbed. Because the thermal gradient is presumed to be initially tangent to the melting curve (Figure 5-9) and the heat flow is from below, the zone of melting will spread downward at a greater rate than it spreads upward. The temperature of rocks below their melting point will be raised more rapidly than that of rocks already at their melting point (Hess, 1960, p. 181) in that the enthalpy of melting is many times the specific heat. The result is a decrease in the thermal gradient below the layer and an increase in the thermal gradient above the layer, as illustrated in Figure 5-9. If the rock is homogeneous laterally, the melt zone will tend to develop in the form of a horizontal sheet. As the proportion of melt increases, convective processes may reduce the difference in rate of advancement of the boundaries. The downward migration of the melt zone will proceed while the melted portion remains close to the solidus. These factors have important implications for the kind of basalt derived from the parental material. Diffusion within the liquid as a result of the developing pressure and temperature gradients will also influence the kind of basaltic liquid.

The development of magma in a temperature and pressure gradient leads to one form of "zoning melting" (Pfann, 1966, p. 254 ff.). The ultimate effect, after partial melting has reached an advanced stage and convection sets in, is to concentrate the more refractory phases containing elements such as Mg and Ca at the high-temperature, high-pressure end of the gradient and the more readily fusible phases containing elements such as Fe and Na at the low-temperature, low-pressure end of the gradient (see Chapter 6). Harris (1957) believed that the potassic basalts in particular are the result of concentrating

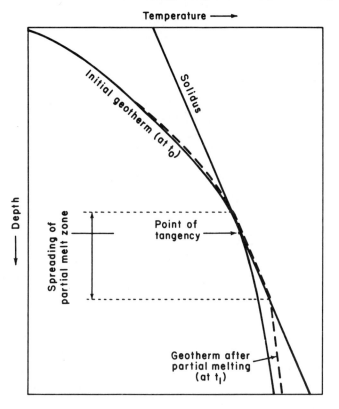

FIGURE 5-9 Preferential spreading of the partial melt zone downward with time ($t_0 \rightarrow t_1$) as a result mainly of the enthalpy of melting. Geothermal gradient increases (in deg/km) above the partial melt zone, is close to the beginning-of-melting curve within the partial melt zone, and decreases below the partial melt zone. After McBirney (1967, p. 31, Figure 9). (With permission of Ferdinand Enke Verlag.)

potassium in a zone-melting process; he pointed out, however, that other expected elements, such as Sr, are not observed to concentrate to the same degree. Vinogradov *et al.* (1971) attributed the major differentiation in the mantle to zone melting on the basis of the Joly–Cotter model presented in Chapter 10.

ADIABATIC RISE

A second mechanism to be analyzed from the standpoint of thermal energy is the generation of magma through the adiabatic rise of a hot

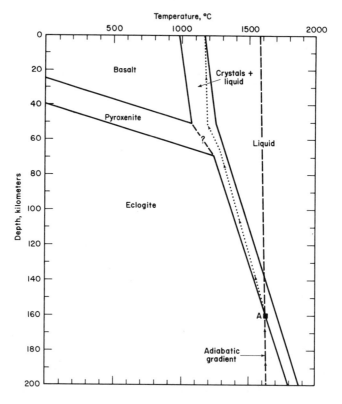

FIGURE 5-10   Eclogite rising adiabatically (dashed curve) begins to melt at 160 km (point *A*). Thermal energy within the mass is consumed for the enthalpy of melting, and the temperature of the partially melted mass changes (dotted line) within the crystal + liquid region. Additional heat is consumed through the pyroxenite region as the remaining eclogite minerals are converted to the phases in basalt. The mass reaches the surface with a temperature at or near the liquidus of basalt. From Yoder (1975a, p. 518, Figure 78). (With permission of the Carnegie Institution of Washington.)

mass (Figure 5-10). It is useful to determine the extent of melting a mass at *A* will undergo as it rises under convective or buoyant forces. If the rise is adiabatic and sufficiently rapid that melting does not take place, the temperature of the mass will decrease slightly according to the equation

$$\frac{dT}{dX}\bigg|_S = \frac{g\alpha T}{C_P}.$$

The adiabatic gradient is about 0.4°C/km (dashed line) when $\alpha = 20 \times 10^{-6}$/deg, $T = 1900°K$, $g = 980$ cm/sec^2, and $C_P = 0.28$ cal/g/deg = 11.7×10^6 dyne-cm/g/deg. The energy within the mass due to superheating is available for the enthalpy of melting, ΔH_m, in an amount equal to

$$H_{T_a} - H_{T_m} = C_P(T_a - T_m),$$

where T_m is the beginning-of-melting temperature and T_a is the adiabatic temperature after rising to a given pressure level. If the average $\Delta H_m = 120$ cal/g and the average effective $C_P = 0.28$ cal/g/deg, eclogite rising from 160 km will arrive at the surface just at or near the liquidus. In the region from 65 to 50 km, reactions take place as those phases of the eclogite facies remaining in the liquid are converted to those of the basalt facies. The reaction may merely be expressed by the change of composition of pyroxene, for example, if garnet has been consumed in the partial melt. The enthalpy of reaction for the conversion of crystalline eclogite to crystalline basalt is endothermic and has a value of about 10.7 cal/g, based on $dT/dP = 145°C$/kbar (Yoder and Tilley, 1962), $\Delta V = 0.051$ cm^3/g ($\rho_{basalt} = 2.85$g/cm^3; $\rho_{eclogite} = 3.33$g/cm^3), and $T = 1275°C$.*

The assumption of ideally adiabatic conditions is, of course, not realistic, nor can it be assumed that crystal settling, differentiation, internal radioactive heating, and other processes have not taken place. An all-liquid magma could be achieved at any stage by separation of the liquid produced, and the composition would be basaltic if the material rising was at or near the eclogite "eutectic." It would be an unlikely event under any circumstances for eclogite to rise, however, because of its high density relative to host rocks such as peridotite. If the material rising were garnet peridotite, the initial liquids presumably would be closely related to basalts in composition; but if that magma, separated at an early stage, rose adiabatically, it would arrive at the surface greatly superheated—also an unlikely event because only very rarely are lavas free from phenocrysts. Heat losses en route or perhaps during storage in an auxiliary chamber may counter that objection.

A more likely event, especially in the midocean environment, is the separation of liquid at relatively shallow depths from peridotite after suitable mineralogical changes from garnet- to spinel- or even plagioclase-bearing varieties. The opportunity for superheating would be minimal and the heat losses greater. The separation of tholeiitic

---

* Oxburgh and Turcotte (1970, p. 1672) estimated the heat of transformation of eclogite to gabbro to be 13 cal/g using the slope of Green and Ringwood (1967).

magma, characteristic of the midocean environment, at shallow depths is consistent with the physicochemical model of Yoder and Tilley (1962) wherein the kind of basalt is dependent for the most part on the depth of separation.

It is evident that the adiabatic rise of a rock at or near its beginning of melting must commence at considerable depths if that is the sole mechanism for achieving complete melting. As will be seen in the next chapter, the loss of crystals en route is apparently necessary to achieve the observed lava compositions, and a shallower point of initiation of the beginning of melting would be more likely. The above discussion serves to emphasize the importance of the enthalpy required for melting.

6 Physicochemical Constraints on Melting

The melting process is constrained by the physicochemical relations of the components of the natural rock systems in accordance with principles that can be illustrated by simpler systems. Although the number of variables in natural processes far exceeds those that can be identified and controlled by experiment, the major variables can be successfully investigated in the laboratory, and a close approach to the natural systems can be achieved.

METHODS OF INITIAL MELT WITHDRAWAL

One of the requirements of a realistic melting process is the generation of large volumes of magma of relatively uniform composition. Estimates of the volumes of lava extruded in some of the great lava floods are given in Table 6-1. On the Columbia River plateau the average rate of extrusion was 0.02 km^3/yr, and on the Deccan plateau the discharge rate was about 0.025 km^3/yr—not excessive rates relative to present-day emissions of about 2 km^3/yr from all 490 active volcanoes plus approximately 5 km^3/yr along midocean ridges. There appear to be four major physicochemical means of achieving such large volumes of relatively homogeneous magma, the first under essentially equilibrium conditions and the others under nonequilibrium conditions: (a) batch melting, (b) fractional melting, (c) zone melting, and (d) disequilibrium melting.

105

TABLE 6-1 Estimates of the Volume and Duration of Large Eruptions

Region	Reference	Estimated Area (km²)	Average Thickness (km)	Volume (km³)	Duration (m.y.)
Karroo lavas, Republic of South Africa, Rhodesia, Botswana, and Lesotho	Cox (1970)	2,000,000	~0.7	1,400,000	20–100
Siberian Plateau, Soviet Union	Lure and Masaitis (1964); Nesterenko and Almukhamedov (1973)	2,500,000	0.36	900,000	130
Parana Plateau, S. Brazil, Paraguay, Argentina, and Uruguay	Cordani and Vandoros (1967)	1,200,000	0.65	780,000	30
Deccan Plateau, Western India	Subramanian and Sahasrabudhe (1964)	500,000	~1	500,000	10–20
North Australia volcanics, Australia	Dunn and Brown (1969)	400,000	~1	400,000	—
Columbia River Plateau, United States	A. C. Waters (in Kuno, 1969)	220,000	0.9	195,000	10

106

BATCH MELTING

The forsterite–diopside–pyrope system at $P = 40$ kbar (Figure 6-1) serves as a useful model to illustrate the first two methods of magma production. Consider the bulk composition X with 60 percent forsterite, and assume the system is ideally ternary. As the temperature is raised, melting begins at 1670°C at the eutectic composition, E. At that temperature, 30 percent liquid is produced, measured by the lever rule.* The residue, R, has the mineralogy of garnet dunite. The liquid produced isothermally can be removed in at least two different ways. It can be removed as one batch or it can be removed as rapidly as it is produced. The rate at which energy is supplied, discussed in Chapter 5, is one of the controlling factors. At any stage during isothermal removal, the liquid composition will be identical with the eutectic composition except for the trace elements.†

FRACTIONAL MELTING

The principles of fractional melting have been presented by Presnall (1969) and Roeder (1974), and, as Bowen (1928, p. 31) pointed out, it is not simply the reverse of fractional crystallization. Consider the removal from the site of generation of a few percent liquid as it is formed.‡ The composition of the remaining rock is no longer X but is depleted of a few percent of the components in composition E, that is, X'. Because X' is in the same three-phase triangle, the new composition also begins to melt at the temperature of E, and the liquid has the

---

* The lever rule gives the proportion of liquid to crystals by a line constructed from the eutectic composition, E, through the bulk composition, X, to its intersection with the join representing the residual phases, R. The proportion of the line from X to R represents the amount of liquid, and the proportion of the line from E to X is the amount of residual phases.

† The definition of a trace element is obscure because under one set of conditions an element may act as a major element (e.g., Fe in fayalite) and under other conditions, behave as an element in trace amounts (e.g., Fe in quartz). It may be useful to restrict the term trace element to a region of behavior of any element that satisfies Henry's law, that is, when its activity is proportional to its concentration.

‡At first glance the removal of a few percent liquid seems to be an unlikely process; that it can occur, however, is abundantly demonstrated by layered intrusions. The development of monomineralic layers at the base of a crystallizing magma must involve similar processes in which the last vestige of interstitial liquid is squeezed or diffused from the cumulate. The same processes apparently are effective on a millimeter scale; for example, Yoder and Tilley (1962, p. 532, Plate 10C), in an experiment on eclogite at high pressures and temperatures, observed an accumulated garnet layer devoid of clinopyroxene that was overlain by quench clinopyroxene.

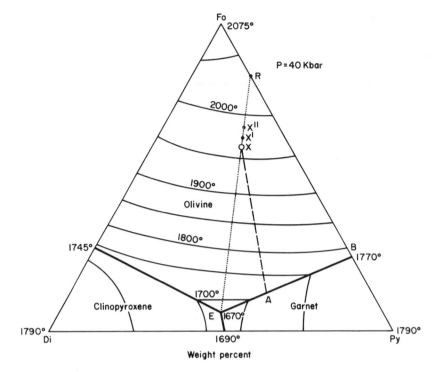

FIGURE 6-1   The forsterite(Fo)–diopside(Di)–pyrope(Py) system at P = 40 kbar (after Davis and Schairer, 1965, p. 124, Figure 35). E is the piercing-point composition. The temperature of the beginning of melting is assumed to be sufficiently close to 1670°C that the behavior is eutecticlike. X is a bulk composition considered analogous to a peridotite, and X' and X'' are successive residual compositions resulting from the fractional melting of X. B is the composition of liquid produced from the fractional melting of R. The dotted line is a construction line to illustrate the lever rule, and the dashed line is the locus of compositions of liquids produced by melting of X at temperatures indicated by the isotherms. (With permission of the Carnegie Institution of Washington.)

composition of E. Remove another few percent liquid, and the composition of the residue becomes X''. Each small batch of liquid has the same composition, until the composition of the residual crystals reaches that of R. The residual composition R, completely exhausted of the Di component, cannot produce any more liquid of the composition E at that temperature, and melting stops. The temperature must be raised to 1770°C before melting can be resumed, and then the liquid will have the composition B. No liquids of intermediate compositions can be derived if the removal of each small batch from the site of generation is complete. In the continuous-withdrawal model, fractional melting would produce two unique batches of liquid having the compositions E and B. The final residue would be dunite—or it too could be melted if the requisite high temperatures could be attained, thereby yielding a third unique magma batch.

SUBSEQUENT MELT WITHDRAWAL

The events following the production of 30 percent liquid in a single batch remaining in contact with its residuum are quite different. Further addition of energy now gradually raises the temperature, and the liquid compositions change along the boundary curve between forsterite and garnet, because all the diopside has been consumed, until the composition A is reached. At this point the lever rule indicates that 46 percent liquid has been produced, and all the pyrope is consumed in the liquid. The liquid compositions now leave the boundary curve along a line constructed through the bulk composition toward forsterite. Complete melting is achieved when the temperature reaches the liquidus of the bulk composition X. A succession of liquids could thereby be produced having a range in composition between the initial and final liquid as long as the residuum remains in equilibrium with the liquid. Liquids withdrawn at any stage of the process would retain their olivine–hypersthene normative character.*

Batch melting and fractional melting produce the same kind of liquid *initially*, but subsequent liquids are quite different. The proportions of forsterite, diopside, and pyrope are sufficiently close to those of natural garnet peridotites to warrant careful consideration of the model system. The norm of the "eutectic" is $Fo_{20}Di_{22}An_{32}En_{26}$, a reasonably close approach to a pyroxene-rich basaltic composition. The model system emphasizes the large amount of liquid of a single composition produced initially and the importance of knowing when and how the liquid is removed from the residuum.

IMPORTANT PRINCIPLES AND QUESTIONS

The simple phase-equilibrium diagram in Figure 6-1 illustrates several other important principles relevant to magma generation and raises some pertinent questions.

---

* Those liquids having compositions to the Py side of a line from Fo to Di_1Py_1 (mol) would also contain Cor, as well as An + En + Fo in the norm. In Figure 6-1, a line from Fo through X and A to the Di–Py side line marks the boundary of corundum-normative liquids. To the left of that line, the norm is Di + An + En + Fo.

1. How does the proportion of phases in the starting material influence the kind and amount of liquid?

2. At what proportion of liquid does the residuum disaggregate?

3. What proportion of liquid is produced in natural peridotite as a function of temperature?

4. Is the melting of natural peridotite eutecticlike?

5. Which phase of garnet peridotite is consumed first in the liquid, clinopyroxene or garnet?

Each question will be addressed in order, and the need for additional experimental data will be evident.

PROPORTION OF PHASES IN PARENTAL MATERIAL

Consider individually in Figure 6-1 a bulk composition of 90 percent forsterite, one of 90 percent clinopyroxene, and one of 90 percent garnet and appropriate amounts of the remaining phases. Each bulk composition will begin to melt at the same temperature, and it will yield an initial batch of liquid of the same composition whether by batch or fractional melting. The main differences arise after the eutectic liquid is attained, depending on which phase is consumed first. As each phase is consumed in batch melting, the liquid changes composition along one of the three boundary curves. The nature of the change in magma composition should therefore reflect the effects of the absent phase. In fractional melting, the loss of a phase—no matter how minor—will cause melting to cease temporarily in the present model. As illustrated in Figure 6-1, an abrupt and discontinuous change of magma composition takes place with each loss of phase. (In cases of complex solid solution, the loss of a phase may cause an abrupt but continuous change in magma composition.)

The above discussion emphasizes the identical nature of the initial liquid irrespective of the proportions of the phases in the starting material. On the other hand, the proportions of the phases will greatly influence the trace element content of successive batches. Because the partitioning of the trace elements into the liquid is not in the proportion of the major phases, one can expect successive batches of liquid to have different trace element contents even though the major element content remains the same. Depletion schemes of trace elements, especially the rare earths (see Chapter 8), on partial melting have been investigated by Schilling and Winchester (1967, p. 277, Figure 9).

The apparent mystery that surrounds trace amounts of elements stems from the drastic changes in partition coefficients at low concentrations. The partition coefficients must be measured at a series of concentrations to determine the species in the melt and the activity coefficients needed to obtain the thermodynamic equilibrium constant. The thermodynamic equilibrium constant is dependent on such intensive parameters as temperature and pressure, but not on the concentrations of the constituents (Mysen, 1976b). The drastic changes in partitioning between crystals or between crystals and liquid are probably due to site restrictions in the competing crystal structures. The common belief that trace elements are not buffered because they are not a required constituent of a phase does not seem to be valid. Partitioning in itself is a buffering process and when calibrated will no doubt yield quantitative information on the processes of magma generation. The observed large variance in the abundances of trace elements gives a considerable advantage in interpretation over the small variance in the abundances of major elements in recording the initial stages of melting. The difficulty in the interpretation of trace element abundances from derived rocks is the same as for major element abundances—the composition of the parental material is not well defined.

The differences in *isotopic* proportions of elements such as strontium between oceanic island and ridge basalts described by Hedge and Peterman (1970) and by Hart *et al.* (1973) cannot be ascribed to fractional crystallization or fractional melting. [No evidence for the fractionation of strontium isotopes in nature has ever been reported (Faure and Powell, 1972, p. 21).] They must, therefore, be due to differences in source, perhaps produced at the time of accretion or as a result of later events. If the earth underwent a melting stage, the long cooling period may have resulted in a continuum of layers or perhaps a random aggregation of diapirs having different rubidium contents or ages of crystallization, or both. The higher $^{87}Sr/^{86}Sr$ in oceanic island basalts relative to ridge basalts would suggest an older source, which presumably is also deeper if crystallization of the earth progressed outwardly from the inside. On the other hand, the difference may reflect a source for oceanic island basalts with a higher rubidium content—that is, a source that has not been depleted of its alkalies. In either case, in the view of Hofmann and Hart (1975), the composition of the mantle appears to be heterogeneous both horizontally and vertically. The Pb isotope content of these basalts also supports the view that the magmas came from sources having separate and independent U/Pb systems (Gast *et al.*, 1964; Tatsumoto, 1966; Sun *et al.*, 1975).

For the above reasons, the writer believes that the upper mantle may be quite heterogeneous in the proportion of phases, but not in the kinds of phases, so that liquids close to basalt in composition will generally be the product of the essentially anhydrous melting of garnet peridotite below about 50 km. The concept of petrographic provinces (Judd, 1886), now well established, suggests that the minor mineral content (e.g., phlogopite, amphibole, apatite) and trace element content vary but that the major mineral assemblage at the site of magma generation is about the same worldwide.

DISAGGREGATION OF PARENTAL MATERIAL

Continued melting will eventually lead to the collapse of the bridging structure of the crystals. No data are available concerning the dis-aggregation of parental material under the conditions of load or stress within the mantle; some limits can be placed on the amount of melting, however, from theory and field observations at the surface. The effective viscosity of a suspension of uniform rigid spheres, based on a relationship provided by Einstein (1906, 1911), was given by Roscoe (1952) as

$$\eta_{eff} = \eta_{fluid}(1 - 1.35C)^{-2.5},$$

where C is the concentration of spheres. When C is 0.446, the effective viscosity has increased one order of magnitude. Without regard for grain shape or change of viscosity with temperature, it appears, there-fore, that collapse of an equigranular rock structure begins to take place when the rock is half melted.

Specimens collected from a drill hole in the lava lake at Makaopuhi, Hawaii, were measured for their glass content after quenching by Wright et al. (1968). The data are plotted in Figure 6-2. The crust formed when the liquid content was reduced to 45 wt%.

These arguments lead to the view that about 45 percent melting would lead to the collapse of the bridging structure of an undisturbed cystalline mass. Thus, it appears that considerable melting would be required to detach a crystal mush as a diapir under undisturbed conditions. The extent of melting that must obtain before detachment under disturbed conditions, the more likely circumstance, has not yet been ascertained.

AMOUNT OF LIQUID VERSUS TEMPERATURE

The amount of liquid produced as a function of temperature in a batch process where the liquid remains in equilibrium with the residuum is

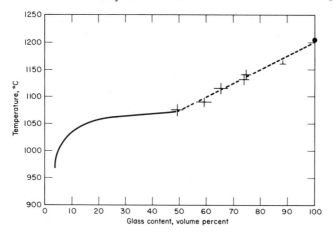

FIGURE 6-2 Glass content of samples collected at various temperatures from Makaopuhi lava lake. Solid line, drill core from crust of lava lake. Crosses, melt collected through drill hole and quenched. Inverted T, pahoehoe toe having maximum temperature measured with an optical pyrometer. Black dot, liquidus temperature of pumice measured in the laboratory. Crust formation occurs when the liquid content is reduced to 50 vol% (= 45 wt%). From Wright *et al.* (1968, p. 3199, Figure 12). (With permission. Copyrighted by the American Geophysical Union.)

shown in Figure 6-3 for the model system forsterite–diopside–pyrope at $P = 40$ kbar. This thermal behavior is compared with the modal data of Scarfe *et al.* (1972, p. 473) obtained by the point-counting technique on the melting of natural garnet peridotite at 1 atm. Spinel instead of garnet was observed at the lowest temperature (1200°C) investigated, where the glass content was "nil." The initial stages of melting are eutecticlike, and an abrupt change in the amount of melting with temperature takes place upon the loss of the clinopyroxene and orthopyroxene.

The melting of a garnet peridotite was also studied at about $P = 20$ kbar by Ito and Kennedy (1967, p. 524, Table 2), who estimated the proportion of glass after quenching. A plot of their data is also given in Figure 6-3. No garnet is reported in any of the runs near 20 kbar, and unfortunately the data are insufficient at pressures where garnet is stable in the presence of liquid. The data of Ito and Kennedy may be interpreted as a continuous smooth curve similar to the liquidus of the simplified forsterite–fayalite system or with cusps where phases are consumed. Although the latter interpretation of their data at $P = 20$

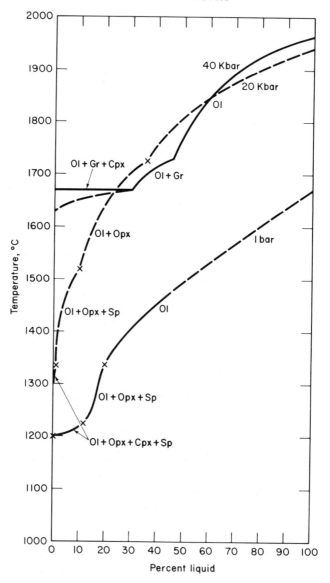

FIGURE 6-3   Percentage of liquid produced on melting to various temperatures. Curve at 1 bar is for a natural garnet peridotite (determined by Scarfe *et al.*, 1972, p. 473, Figure 1). Curve at 20 kbar was constructed from the data of Ito and Kennedy (1967, p. 524, Table 2) on a natural garnet peridotite. Curve at 40 kbar is deduced from Figure 6-1: solid curve assumes eutectic behavior, and optional dashed branch of curve illustrates quaternary melting behavior.

kbar is preferred, the melting is not eutecticlike. Melting related to a continuous series of solid solutions is implied. The few data obtained by Ito and Kennedy at $P = 40$ kbar suggest a return to eutecticlike melting where almost 20 percent liquid appears as a result of the consumption of both clinopyroxene and garnet in a narrow temperature interval.

It is difficult to determine the amount of glass in an experimental run; nevertheless, such data are most useful in reconstructing the physicochemical relations for a natural rock. Additional experiments for the purpose of accurately determining the proportion of liquid as a function of temperature and pressure are clearly needed. The use of radioactive tracers, beta-track counting, and the scanning electron microscope may aid in collecting these data. In this way it may be possible to reconstruct the phase diagrams applicable to natural rocks under conditions similar to those at their site of generation. (It is perhaps not out of place to encourage the prospective experimenter to choose his natural specimens or synthetic mixtures with due regard for their characteristics as representative of mantle material that will yield liquids of basaltic composition, i.e., undepleted.)

Because of the great importance the writer attaches to the eutectic and eutecticlike behavior of major basaltic magma types, the next chapter is devoted to the supporting arguments for questions 4 and 5 above.

ZONE MELTING

A nonequilibrium process for accumulation of magma related to fractional crystallization was suggested by Harris (1957) and by Shimazu (1959), based on the principles of zone melting outlined by Pfann (1952, 1966). A liquid layer or partially melted layer is presumed to move upward in the mantle by simultaneous melting and mixing of the roof rocks and crystallization and deposition at the floor of the magma. It is assumed that there is relatively little heat loss until near-surface conditions are attained, the mantle is relatively homogeneous, there is no diminution in volume of liquid, and the temperature of the beginning of melting of the rocks encountered decreases with decrease in pressure. The melting of the cooler roof rocks is accomplished through convection (*cf.* convective heat transfer models, Chapter 10) by adiabatic heat transfer from the hotter base of the molten layer. The enthalpy of melting required for melting the roof would be slightly less than the enthalpy of crystallization released at the floor because of the pressure effect. The effectiveness of the heat transfer is governed by the permeability, if any, and the thermal conductivity of the roof rocks. A newly formed magma layer would begin to move upward as soon as the proportion and total volume of melt had reached a sufficient amount for convection to become effective.

The unique aspect of zone melting is the amount and composition of liquid that is transferred as the assimilation–deposition process advances. Consider a partially molten zone 10 km thick (Figure 6-4A), the top of which is at 200 km depth in a homogeneous upper mantle of garnet wehrlite. For simplicity, it will be assumed that the garnet wehrlite consists of $Fo_{60}Di_{20}Py_{20}$ so that the composition lies simply on the line joining Fo and the eutectic composition at 40 kbar (see Figures 6-1 and 6-5). In addition, it is assumed that the phase relations of the system remain unchanged over the pressure interval considered—a most unlikely circumstance. As the melt zone at T_0 moves a small distance (dz) upward, forsterite is precipitated at the base and with the liquid ℓ is mixed the new partially melted layer, dz in thickness and having the bulk composition of the initial material Y. With thorough mixing, dz is incorporated into the mush, whose bulk composition becomes Y' as forsterite precipitates. Because forsterite is being removed and only Y added, the liquid ℓ increases in amount as indicated by

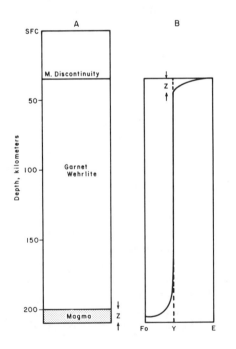

FIGURE 6-4 *A*. Schematic section through crust and upper mantle, assumed for simplicity to be of garnet wehrlite composition, showing the position of a newly formed partially melted zone (Z) prior to the onset of convection and subsequent rise through the mantle.

B. Composition-versus-depth diagram for the simplified system Fo–$Cpx_{50}Gr_{50}$ having a eutectic composition E. The mantle has the composition Y (dashed line) prior to zone melting and variable composition (solid line) after a single pass of zone melting.

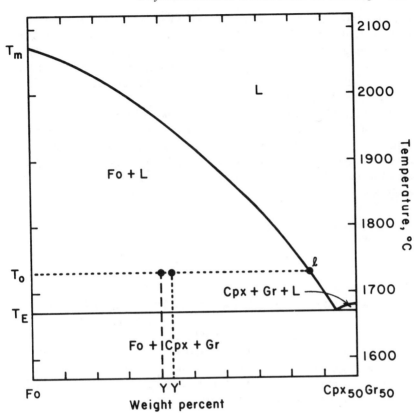

FIGURE 6-5 Schematic T–X diagram for the system Fo–$Cpx_{50}Gr_{50}$ based on the Fo–Di–Py system of Davis and Schairer (1965, p. 124, Figure 35) for $P = 40$ kbar. T_0 is the initial temperature of the partially melted zone rising through the mantle (Figure 6-4A) by zone melting. ℓ is the composition of liquid in equilibrium with forsterite at T. Y is the initial composition of the melting zone at the onset of zone melting. Y' is the composition of the advanced partially melted zone resulting from the mixing of about 7 percent liquid ℓ with the initial partially melted zone.

the lever rule. The process of melting and crystallization proceeds with continued increase in volume of the liquid ℓ.

In the meantime, below the base of the displaced melt layer, the composition of the liquid must change accordingly by precipitating forsterite. With continued movement upward of the molten zone the liquid cooling at the base approaches the composition E (Figure 6-4B) by crystallizing forsterite. At the temperature of E the eutectic liquid crystallizes all the participating phases in the proportions of the initial bulk composition Y. Therefore, when the base of the molten layer reaches a depth of about 170 km the composition of the material crystallizing out at the temperature of E will be exactly equal to that of the material being melted at the top of the molten layer; that is, Y, the original bulk composition. No further enrichment of the liquid is achieved until the last melt zone is reached.

When the heat losses prohibit the advance of the molten layer, the accumulated liquid crystallizes, presumably just below the M discontinuity. The last portion of the zone to crystallize has the composition of the eutectic. The width of the zone of enrichment near the surface is equal to the width of the initial melt zone. Successive waves of later melt zones would increase the width of the zone of enrichment.

Zone melting would eventually produce a melt of the eutectic composition equal in volume to that produced by batch melting or fractional melting. Unfortunately, the process would be terminated if a refractory layer was encountered. Furthermore, the initiation of the rise of a zone requires melting to an extent considerably greater than that envisaged in the mantle. For maximum efficiency, the rise of the molten zone must take place at a rate that ensures complete mixing in the liquid and complete removal of the crystals from reaction with the liquid. The close approach to equilibrium attained within the times available for magma generation and movement appears to relegate the zone-melting process to a minor role. The life of a magma zone 20 km in thickness generated at 100 km in the upper mantle is of the order of 10 m.y., according to Shimazu (1961, p. 193).

DISEQUILIBRIUM MELTING

O'Nions and Pankhurst (1974, p. 624 ff.) suggested that the constituent minerals of garnet peridotite enter the melt in proportions other than those determined by equilibrium melting relations. This model was proposed to account for the variations of strontium isotopes in rocks obtained from the mid-Atlantic ridge. Because the minerals are assumed to contain different $^{87}Sr/^{86}Sr$, the rate at which they contribute to the magma would affect that ratio in the magma and its subsequent fractionation products. O'Nions and Pankhurst attributed the differences in $^{87}Sr/^{86}Sr$ observed in rocks derived from what appeared to them to be a major single magma source under the mid-Atlantic ridge to disequilibrium melting.

The rapid response of garnet peridotite and eclogite of a wide range of grain size to melting conditions at high pressures in the laboratory would rule out such disequilibrium melting. Only one common igneous-rock-forming mineral that melts with great difficulty under anhydrous conditions is known to the writer; that is albite, and it is not anticipated as a phase in rocks of normal basalt composition at high pressures. If anhydrous albite diabases are indeed an igneous product, then one might consider them as a possible candidate for disequilibrium melting, but even then equilibrium melting may be achieved on the order of several months. There is no laboratory experience yet on the melting rate of large megacrysts now being recovered with the eclogite and garnet peridotite nodules from kimberlite pipes. Nevertheless, using measured diffusion coefficients, a grain 1 cm in diameter in a partially molten rock would equilibrate by volume diffusion with a melt in less than a million years (Hofmann and Hart, 1975).

7 Composition of Major Basaltic Magmas

FRAMEWORK OF IGNEOUS PETROLOGY

The key to the most abundant basaltic magma compositions is the composition of the invariant points and related univariant curves controlling the physicochemical behavior of the natural chemical systems at each pressure level where magma separation takes place. The fact that basalts are restricted in mineral content and composition worldwide indicates that they are not random products of the various processes described. It is most unlikely that a specific basalt composition occurs because of the complete melting of a pre-existing mass of exactly the desired composition or is always the product of a specific degree of melting. The repetitious extrusion of tholeiitic basalt, for example, in many successive flows, especially in the ocean environment, cannot depend on specific flow rates of magma, heat losses, or proportion of crystals to liquid. There appears to be a memory to magma generation so that flow after flow of similar composition is extruded over the world and through large intervals of time. The principal control on the restriction of magma types is believed to be the pressure–temperature-dependent phase relations.

BASALT TETRAHEDRON

The concept of a simple model for a wide range of basalt types (Yoder and Tilley, 1957, pp. 156–161; 1961, pp. 106–113) arose from a detailed experimental study of natural basalts and simple systems of principal end-member minerals that make up basalt. The investigation of a number of relatively simple systems combining the early-crystallizing minerals of rocks with late-crystallizing alkali aluminosilicates had established the nature of residual liquids from the crystallization of relatively complex magmas. The diversity of rocks produced in the late stages of crystallization appears to have evolved from very similar source magmas. The basalts, representing these source magmas, consist primarily of feldspar and clinopyroxene with small amounts of one or more of the minerals olivine, orthopyroxene, nepheline, quartz, and iron ores. It was important, then, to investigate systems that deal with these principal phases. Because of the large number of components involved, it was expedient to limit the composition of the phases in the preliminary studies. For experimental simplicity, the iron-bearing end members of these phases were temporarily set aside. The small amounts of K_2O, TiO_2, and MnO found in the basaltic magmas were neglected as well, however, without losing significant phases. If these limitations are applied, the experimental major basalts can be represented for the most part in a normative tetrahedron comprised of larnite(La)–nepheline(Ne)–forsterite(Fo)–quartz(Qz), as displayed in Figure 7-1.

Experimental data on the principal planes were collected mainly by Schairer and Yoder (1964, pp. 65–74) at 1 atm (Table 7-1). The system involves five oxides (Na_2O–CaO–MgO–Al_2O_3–SiO_2), but a close approximation to the phase relations can be made by reducing the effective components to four. An expanded view of the subtetrahedra is given in Figure 7-2. The relationship of each of these subtetrahedra to basaltic magma types is given below.

An analogous tetrahedron based on anorthite can also be constructed (Table 7-2); however, contiguous subtetrahedra are not necessarily compatible where nepheline is a component (*cf.* Schairer *et al.*, 1968, p. 467, Figure 67). A consistent set of relations can be obtained if, in addition, nepheline is replaced by the calcium-Tschermak's molecule $CaAl_2SiO_6$, and the tetrahedron reduces to a portion of CaO–MgO–Al_2O_3–SiO_2 (Schairer and Yoder, 1970). An evaluation of the effect of plagioclase in place of albite in the expanded basalt tetrahedron may be gained by examining those systems relating albite and anorthite to the other end-member components (Table 7-3).

ILLUSTRATION OF PHASE EQUILIBRIA

On the basis of many detailed experiments (Table 7-1), the compositions of liquids in equilibrium with one or more crystalline phases were

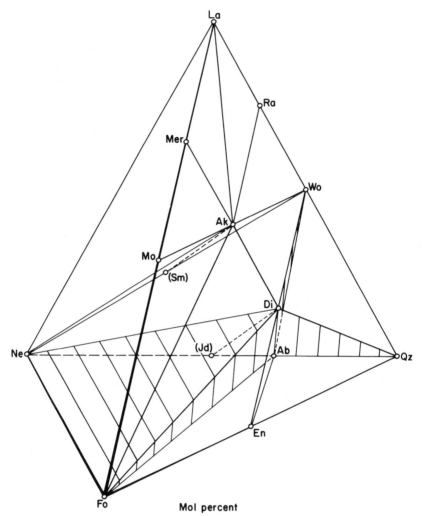

Mol percent

FIGURE 7-1   Normative basalt tetrahedron of Yoder and Tilley (1962, p. 350, Figure 1) expanded by Schairer and Yoder (1964, p. 65, Figure 1) to include melilite-bearing assemblages. Ruled planes are close to, but because of solid solution, not exactly the thermal divides (see flow sheet of Figure 7-6) determined by experiment.

Ab = albite	*Fo* = forsterite	*Mer* = merwinite	*Ra* = rankinite
Ak = akermanite	*Jd* = jadeite	*Mo* = monticellite	*Sm* = soda melilite
Di = diopside	*La* = larnite	*Ne* = nepheline	*Wo* = wollastonite
En = enstatite	*Qz* = quartz		

Related end-member minerals in parentheses are not stable at 1 atm. (With permission of Oxford University Press and the Carnegie Institution of Washington.)

TABLE 7-1 Systems within the Expanded Basalt Tetrahedron Studied at 1 atm, Including Joins Stable Only at High Pressure

System[a]	References
Ab–Ak–Di	Schairer and Yoder (1964, p. 69, Fig. 5)
Ab–Ak–Ne	Schairer and Yoder (1964, p. 70, Fig. 6)
Ab–Ak–Wo	Schairer and Yoder (1964, p. 68, Fig. 4)
Ab–Di–En	Schairer and Morimoto (1959, p. 115, Fig. 16)
Ab–Di–Fo	Schairer and Morimoto (1958, p. 213, Fig. 24)
Ab–Di–Ne	Schairer and Yoder (1960a, p. 278, Fig. 2)
Ab–Di–Wo	Yoshiki and Yoshida (1952, p. 167, Fig. 1)
Ab–Di–Qz	Schairer and Yoder (1960a, p. 278, Fig. 2)
Ab–En–Fo	Schairer and Yoder (1961, p. 142, Fig. 35)
Ab–En–Qz	Schairer and Yoder (1961, p. 142, Fig. 35)
Ab–Fo–Ne	Schairer and Yoder (1961, p. 142, Fig. 35)
Ab–Ne–Wo	Foster (1942, p. 165, Fig. 6)
Ab–Qz–Wo	Not done (two-component joins done)
Ak–Di–Fo	Ferguson and Merwin (1919, p. 109, Fig. 12)
Ak–Di–Ne	Schairer and Yoder (1964, p. 66, Fig. 2); Onuma and Yagi (1967, p. 238, Fig. 4)
Ak–Di–Wo	Ferguson and Merwin (1919, p. 109, Fig. 12); Schairer and Bowen (1942, p. 730, Fig. 5; p. 741, Fig. 6)
Ak–Fo–Ne	Not done (two-component joins done)
Ak–Ne–Wo	Schairer and Yoder (1964, p. 67, Fig. 3)
Ak–Sm	Schairer *et al.* (1965, p. 96, Fig. 17); Yoder (1973, p. 154, Fig. 11)
Di–En–Fo	Kushiro (1972a, p. 1263, Fig. 1)
Di–En–Qz	Kushiro (1972a, p. 1263, Fig. 1)
Di–Fo–Ne	Schairer and Yoder (1960b, p. 70, Fig. 18)
Di–Jd	Schairer and Yoder (1960a, p. 278, Fig. 2)
Di–Ne–Wo	Schairer and Yoder (1964, p. 71, Fig. 7)
Di–Qz–Wo	Ferguson and Merwin (1919, p. 109, Fig. 12)
Fo–Jd	Mao and Schairer (1970, pp. 221–222)

[a]Compositions of the components, arranged alphabetically (symbols as in caption of Figure 7-1):

Ab:	$NaAlSi_3O_8$	Jd:	$NaAlSi_2O_6$
Ak:	$Ca_2MgSi_2O_7$	Ne:	$NaAlSiO_4$
Di:	$CaMgSi_2O_6$	Qz:	SiO_2
En:	$MgSiO_3$	Sm:	$NaCaAlSi_2O_7$
Fo:	Mg_2SiO_4	Wo:	$CaSiO_3$

determined. To illustrate the technique for determining the composition and course of liquids, the Ab–Di–En–Qz subtetrahedron will be used. Polymorphic changes will be ignored for simplicity. Because of plagioclase and pyroxene solid solutions, the system is in fact quinary, but it can be treated for all practical purposes as quaternary. The subtetrahedron can be laid out from the Di apex, and the ternary phase-equilibrium data displayed (Figure 7-3). The ternary invariant points are labeled *a–f*. The "quaternary" boundary curves and the

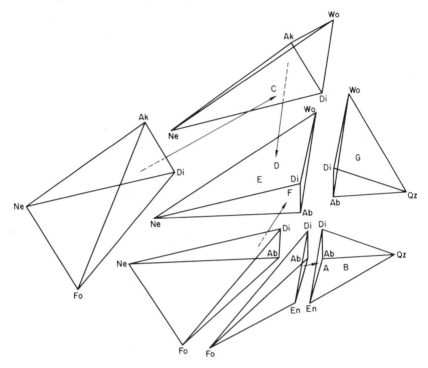

FIGURE 7-2 Subsolidus tetrahedra within the expanded basalt tetrahedra as determined at 1 atm by experiment. Arrows indicate course of liquids to determined invariant points lettered as in Figure 7-6. Because of solid solutions in clinopyroxene, B may be a reaction point with liquids tending toward G.

invariant points A and B are presented in a generalized diagram (Figure 7-4). Because the "quaternary" invariant point B, where Pl, Qz, En_{ss}*, and Di_{ss} coexist with liquid, lies within the subtetrahedron representing the four crystalline phases, it is eutectic. On the other hand, "quaternary" invariant point A, where Fo, Pl, En_{ss}, and Di_{ss} coexist with liquid, does not lie within the subtetrahedron representing its four crystalline phases; hence, it is a reaction point. The four crystalline phases in equilibrium with liquid at point A form the contiguous subtetrahedron (see Figure 7-2). Whereas all proportions of the phases in the subtetrahedron Ab–Di–En–Qz begin to melt at the "quaternary" invariant point B, all proportions of phases in the adjoining subtetrahedron Ab–Di–En–Fo begin to melt at the "quaternary" reaction point A, the

*The subscript ss refers to solid solutions of the designated phase.

TABLE 7-2   Additional Systems Studied at 1 atm, Required to Construct an Expanded Basalt Tetrahedron Based on Anorthite.

System	References
Ak–An–Di	de Wys and Foster (1958, p. 741, Fig. 2)
Ak–An–Fo	Yang *et al.* (1972, p. 167, Fig. 4)
Ak–An–Ne	Not done (two-component joins done)
Ak–An–Wo	Rankin and Wright (1915, p. 40, Fig. 8)
An–Di–En	Hytönen and Schairer (1960, p. 71, Fig. 19)
An–Di–Fo	Osborn and Tait (1952, p. 419, Fig. 5)
An–Di–Ne	Schairer *et al.* (1968, p. 469, Fig. 69)
An–Di–Qz	Clark *et al.* (1962, p. 66, Fig. 6)
An–Di–Wo	Osborn (1942, p. 761, Fig. 5)
An–En–Fo	Andersen (1915, p. 440, Fig. 10)
An–En–Qz	Andersen (1915, p. 440, Fig. 10)
An–Fo–Ne	Not done (two-component joins done)
An–Jd	Mao and Schairer (1970, p. 221–222)
An–Ne–Wo	Gummer (1943, p. 515, Fig. 7)
An–Qz–Wo	Rankin and Wright (1915, p. 40, Fig. 8)

composition of which lies outside the subtetrahedron representing the participating phases. Because of the complexity of such four-component liquidus diagrams and the desirability of examining the results in conjunction with those of several contiguous tetrahedra, a scheme was devised to display the course of liquids in a simple way.

FLOW SHEET

The ternary and quaternary invariant points can be displayed in two dimensions without regard to spatial orientation, as in Figure 7-5. The lowercase letters represent the assemblages in Figure 7-3, and the

TABLE 7-3   Systems Relating Albite and Anorthite to Other End-Member Components of the Expanded Basalt Tetrahedron

System	References
Ab–An–Ak	Schairer and Yoder (1969, p. 104, Fig. 15)
Ab–An–Di	Bowen (1915, p. 178, Fig. 12); Kushiro (1972a, p. 1263, Fig. 1)
Ab–An–En	Not done (two-component joins done)
Ab–An–Fo	Schairer and Yoder (1967, p. 206, Fig. 2)
Ab–An–Ne	Schairer (1957, p. 232, Fig. 35); see also Yoder (1968b, p. 477, Fig. 77)
Ab–An–Qz	Schairer (1957, p. 232, Fig. 35); see also Yoder (1968b, p. 477, Fig. 77)
Ab–An–Wo	J. S. Griffith (Ph.D. thesis, University of Chicago, unpublished)

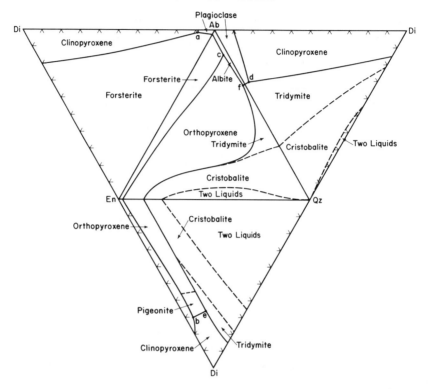

FIGURE 7-3   The system diopside(Di)–albite(Ab)–enstatite(En)–quartz(Qz) laid out to show liquidus relations of the boundary systems at 1 atm. The piercing points and invariant points are labeled with lowercase letters and also appear in Figures 7-4 and 7-5. The sources of the data are given in Table 7-1.

uppercase letters, those in Figure 7-4. The direction of falling temperature on the univariant curves connecting the invariant points is shown by arrows. For this reason, the diagram in Figure 7-5 is described as a flow sheet (Schairer, 1942, 1954).

On the basis of the vast amount of data collected for all the subtetrahedra in Figure 7-1, a flow sheet can be constructed for the petrologically relevant part of the normative tetrahedron. This flow sheet, incorporating the example, is given as Figure 7-6. It is seen that the more primitive liquids, those obtained at the highest beginning-of-melting temperatures, originate from the assemblages Ol–Di–En–Ab, Ol–Di–Ab–Ne, and Ol–Di–Ak–Ne. The liquids generated at the beginning of melting have compositions that lie outside their respective subtetrahedra! They are all reaction points and involve reactions of olivine with liquid. (The importance of this observation is detailed in Chapter 8.) The arrowheads in Figure 7-2 indicate the appropriate subtetrahedron in which the invariant liquid composition lies for crystalline phases in the contiguous subtetrahedron at the arrow tails. The rock names relating to the mineral assemblages are presented in an identical flow sheet in Figure 7-7, corresponding to the univariant lines and invariant points of Figure 7-6.

Olivine tholeiite (*A*), represented by the assemblage Fo–Di–En–Ab, passes by reaction of olivine with liquid through hypersthene basalt and eventually to tholeiite (*B*). This sequence of magmas is separated from the nepheline basanite (*F*) sequence by a thermal maximum or

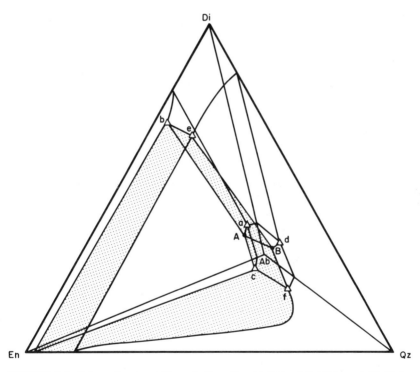

FIGURE 7-4 Schematic assembly of the diopside(Di)–albite(Ab)–enstatite(En)–quartz(Qz) system shown in Figure 7-3 to illustrate the probable location of the "quaternary" invariant points, *A* and *B*. Because of the plagioclase and solid solutions in pyroxenes that lie outside the tetrahedron, the system is quaternary only to a first approximation. Polymorphism is neglected for simplicity. "Ternary" invariant points are marked with triangles, and "quaternary" invariant points with black dots. Some of the sides of the orthopyroxene volume are stippled to aid in visualizing the three-dimensional relations.

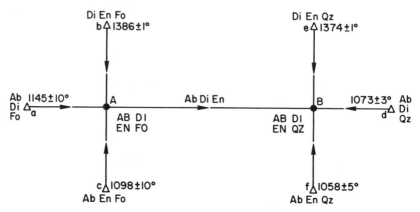

FIGURE 7-5   Flow sheet of the "univariant" and "invariant" liquids in the system diopside(Di)–albite(Ab)–enstatite(En)–quartz(Qz) displayed in Figures 7-3 and 7-4. Lowercase letters represent "invariant" points in Figure 7-3; uppercase letters, those in Figure 7-4. Positions of "invariant" points are similar to those in Figure 7-4; however, no other spatial relations are to be inferred. Arrows indicate direction of falling temperatures. The temperature of A must be $< 1098° \pm 10°C > B$, and the temperature of B must be $< 1058 \pm 5°C$.

divide at 1 atm (Yoder and Tilley, 1962, p. 398) represented by the olivine basalts bearing minor occult nepheline or hypersthene. The nepheline basanites pass by reaction of olivine with liquid through nepheline tephrite and eventually to wollastonite–nepheline tephrite (E). This sequence of magma is separated from the olivine–melilite nephelinites by another thermal maximum represented by olivine nephelinites. The olivine–melilite nephelinites (C) pass by reaction of olivine with liquid through melilite nephelinite and wollastonite–melilite nephelinite (D), and by reaction of melilite with liquid eventually to wollastonite–nepheline tephrite (E).

"PARENTAL" MAGMAS

It would appear that three "parental" magmas (olivine tholeiite, nepheline basanite, and olivine–melilite nephelinite) are required to derive the magma sequences observed at 1 atm. As will be shown below, these "parental" magmas can be derived from a single parental material by various means involving high pressure with or without volatiles. The important observation, and the main purpose in presenting this detailed argument, is that the major magma types observed at

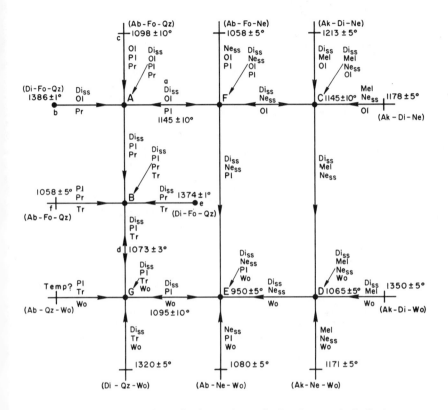

FIGURE 7-6   Flow sheet determined experimentally for the petrologically important portions of the expanded basalt tetrahedron in Figure 7-1 (after Schairer and Yoder, 1964, p. 72, Figure 8; 1967, p. 209, Figure 5). Abbreviations as in Figure 7-1; also, *Mel* = melilite; *Pl* = plagioclase; *Pr* = protoenstatite; *Tr* = tridymite; *ss* = solid solution. Uppercase letters are quaternary "invariant" points. Lowercase letters are ternary "invariant" points from Figure 7-5. (With permission of the Carnegie Institution of Washington.)

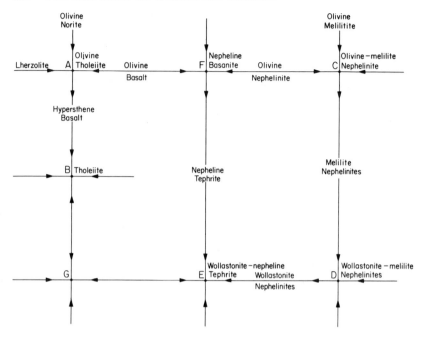

FIGURE 7-7   Rock nomenclature diagram corresponding to the flow sheet in Figure 7-6. For the most part, names apply to extrusive rock types. Invariant point *G* may be represented in nature by some metamorphic rocks. The reaction points *A*, *F*, and *C* involve loss of olivine. From Schairer and Yoder (1964, p. 73, Figure 9). (With permission of the Carnegie Institution of Washington.)

the surface of the earth are closely* represented by "invariant"† points or "univariant" curves. They obviously are derivative liquids controlled mainly by the phase-equilibrium relations. It is concluded that:

*The normative proportions of Cenozoic basalts determined by Chayes (1972, p. 293, Figure 4) support this thesis. It is important to document these conclusions further with modes of basalt in the same manner as has been done with granites (Tuttle and Bowen, 1958, p. 75, Figure 38). The electron microscope is now a convenient tool for such a tedious but rewarding investigation of the fine-grained basalts.

†The assemblages labeled as "invariant" actually melt over a small range of temperature. Such points are not accurately determined because of the difficulty of observing the first appearance of liquid in a four-phase assemblage. To the best of the writer's knowledge, the greatest range of melting at the "invariant" points is displayed by *C*, for which Schairer et al. (1965, p. 97, Figure 29) observed a 40°C interval. That is, $Di_{ss} + Ol + Mel + Ne_{ss}$ coexisted with liquid as a univariant assemblage throughout that temperature interval.

1. Igneous rocks are not random products of the melting of varying mineral assemblages. In short, the flow sheet is the fabric of igneous petrology and demonstrates that there is a rigorous control on the generation of basaltic and nephelinic magmas.

2. The magmas that arrive at the surface have been equilibrated for the most part at successive levels on the way to the surface or have achieved equilibrium at near-surface conditions, either along conduits or in auxiliary reservoirs.

3. The small number of invariant points at 1 atm involving olivine in the 5-component system described, as well as in the systems K_2O–MgO–Al_2O_3–SiO_2 (Schairer, 1954), CaO–FeO–Al_2O_3–SiO_2 (Schairer, 1942), CaO–MgO–Al_2O_3–SiO_2 (Schairer and Yoder, 1970, p. 212), and Ne–Di–An–Fo–Ak (Schairer *et al.*, 1968, p. 470), is a strong argument for believing in a single unique assemblage, or at least not more than two, as potential parental material for the generation of basalts. In other words, in spite of the great range of compositions studied, the number of beginning-of-melting points is exceptionally limited, and these points, based on the abundance of basalt, are presumed to be close in composition to the liquid(s) that yields basalt.

NEED FOR FLOW SHEET FOR HIGH PRESSURES

There is a clear need to ascertain the flow sheet for the same array of bulk compositions at a series of pressures such as $P = 10, 20, 30,$ and 40 kbar.* Only a few data are available, however, and there is as yet no method for dealing adequately with large ranges of solid solution. (The system Na_2O–CaO–Al_2O_3–SiO_2 can be illustrated by a flow sheet only if it is assumed that albite and anorthite are separate, independent phases—an obvious example of the shortcomings of the flow sheet for systems involving continuous solid solution.) The direction of shift in composition of some of the invariant points with pressure has been indicated. For example, the invariant point for the assemblage Fo–En–Ab (on the base of the iron-free normative basalt tetrahedron) shifts with increasing pressure from quartz-normative through hypersthene-normative to nepheline-normative compositions (Figure

---

*There are no theoretical reasons for choosing these pressure levels except that important reactions in the synthetic and natural systems investigated occur near them. Because of the advantages of correlating the work of many laboratories around the world, it is hoped that these pressure levels, within the convenient working range for various apparatus available today, will be found acceptable to the majority of investigators. The exciting opportunities for research under conditions up to 1 Mbar and 2500°C, attainable in the diamond-anvil, laser-heated pressure cell (Bell and Mao, 1975; Mao and Bell, 1976), will no doubt lead to other preferred pressure levels as the important reactions are identified.

7-8). A single bulk composition could therefore produce a wide array of basaltic liquids, as suggested by Yoder and Tilley (1961, 1962), provided the liquids were separated at different pressures. In general, it appears that the nepheline-normative basaltic magmas are derived at greater depths than the hypersthene or quartz-normative magmas.

Volatiles produce similar effects, depending on their composition. For example, Kushiro (1972b, p. 316, Figure 3) demonstrated that at $P = 20$ kbar, H_2O shifts the same invariant point described above to more siliceous compositions, whereas Eggler (1974, p. 216, Figure 1) showed that CO_2 shifts the composition of that invariant point to more nephelinitic compositions (Figure 7-9).

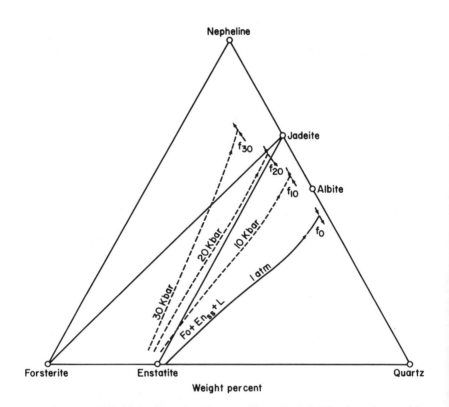

FIGURE 7-8   Shift of the forsterite (Fo)–enstatite$_{ss}$ (En$_{ss}$) liquidus boundary and invariant points (f) with pressure in the forsterite–nepheline–quartz system. Data at 1 atm (solid line; f_0) from Schairer and Yoder (1961, p. 142, Figure 35). Estimates (dashed lines) at 10, 20, and 30 kbar from Kushiro (1968, p. 625, Figure 4). Tie lines are illustrated for 30 kbar. (With permission of the Carnegie Institution of Washington and the American Geophysical Union.)

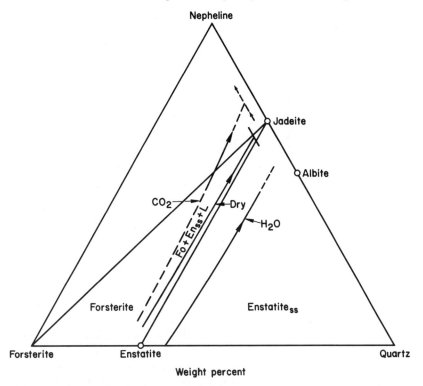

FIGURE 7-9   Shift of the forsterite (Fo)–enstatite$_{ss}$ (En$_{ss}$) liquidus boundary and invariant points with H_2O and CO_2 from volatile-absent (Dry) positions at 20 kbar in the forsterite–nepheline–quartz system. From Eggler (1974, p. 216, Figure 1). (With permission of the Carnegie Institution of Washington.)

MAGMA SEPARATION

It is essential, therefore, to specify not only the depth of separation of magma but also the amount and kinds of volatiles present in the magma in order to fix the probable magma type and eventual trend of derivative magmas. It is not useful to attempt at this time to outline the possible flow sheets at various pressures without considerably more data than are now available. Comprehensive studies at high pressures are now under way, and a new array of problems is emerging. One of these problems involves the presence or absence of the key phase olivine, and because of its importance full discussion is given in the next chapter.

8 Olivine in the Parental Material

OLIVINE FROM INITIAL MELT

In Chapter 2 arguments were presented for believing that olivine is the dominant phase in the parental material of basaltic magmas. If olivine is in the parental assemblage and the system is eutecticlike, then liquids removed at the beginning of melting should reprecipitate olivine on cooling. The liquid removed from the eutectic at 1 atm of the forsterite–diopside–anorthite system, presented in Figure 2-3, would crystallize olivine as well as anorthite and diopside. Similarly, in the forsterite–diopside–pyrope system presented in Figure 5-6, the liquid removed from the "eutectic" at 40 kbar would crystallize olivine as well as clinopyroxene and garnet. If, on the other hand, the system is at a pressure where the beginning of melting is a reaction point in which olivine is the phase reacting with liquid, then the liquid removed from that reaction point will not crystallize olivine at temperatures below that of the reaction point. Furthermore, the absence of olivine in the beginning-of-melting assemblage at one pressure does not preclude its appearance from similar bulk compositions at another pressure.

ABSENCE OF OLIVINE IN BASALTS AT HIGH PRESSURES

In the light of these principles, the results obtained on the reported behavior of natural basalts at the high pressures where they are presumed to have originated or separated from their parental material are surprising. The melting relations of an olivine tholeiite were studied up to 30 kbar by Thompson (1972, p. 409, Figure 24), who found the liquidus phase to be clinopyroxene above 10 kbar and garnet above 32 kbar (Figure 8-1). No olivine was found in any of the assemblages at any temperature at pressures above 13 kbar! In another olivine tholeiite, investigated by Cohen *et al.* (1967, p. 496, Figure 4), the liquidus phase was clinopyroxene above 10 kbar and garnet above about 28 kbar (Figure 8-2). No olivine was observed in any of the assemblages above approximately 10 kbar! A nepheline-normative olivine basanite subjected to a range of conditions by Arculus (1975, p. 514, Figure 74*A*) had olivine on the liquidus up to only 21 kbar (Figure 8-3). The highest pressure at which olivine was observed in any assemblage was 26 kbar! Other less complete studies support the observation that olivine is absent in assemblages obtained from natural basalts at high pressures. Eclogites, the high-pressure equivalent of basalt according to Yoder and Tilley (1962), also fail to produce olivine at high pressures.

If olivine is an essential liquidus phase in the generation and fractionation of basaltic magma, then the above data would appear to put some severe constraints on its depth of origin or separation. The olivine tholeiite from New Mexico would have to originate at depths less than 30 km at temperatures between 1200° and 1250°C. Separation of an olivine tholeiite liquid might be appropriate under such conditions in the midocean ridge environment, but its generation at 30 km is not considered a likely event under continental or abyssal oceanic areas. The basanite from Australia would have to originate at depths less than 75 km at temperatures between 1225° and 1325°C. It would be difficult to explain a genetic relationship of eclogite and garnet peridotite nodules to their host basanites and alkali basalts if the depth of origin was limited to 75 km. The nodules could be unrelated to the basaltic liquids and merely samples of shallower and cooler parts of the mantle ripped off as the magma erupted. The high temperatures and high pressures deduced from the chemical composition of coexisting minerals in the nodules and the chemical relationships between the nodules and the lavas do not support that view (Beeson and Jackson, 1970).

At least three other possible conclusions may be drawn from these observations on basalts and eclogite:

1. Basalts and eclogite are not derived from an olivine-bearing parent.
2. Basalt magmas are altered in composition on the way to the surface.

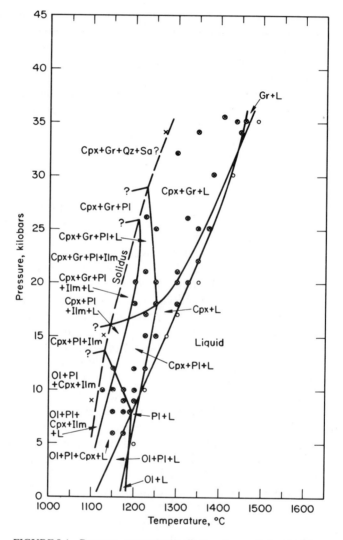

FIGURE 8-1   Pressure–temperature diagram for an olivine tholeiite from Snake River, Idaho, determined by Thompson (1972, p. 409, Figure 24). (With permission of the Carnegie Institution of Washington.)

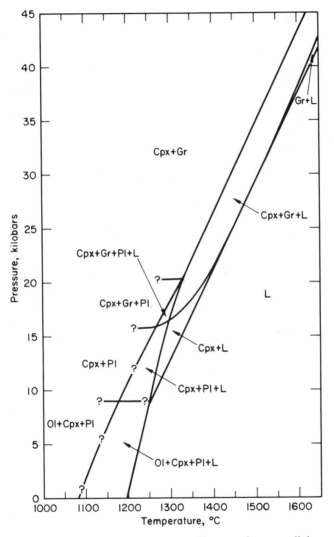

FIGURE 8-2 Pressure–temperature diagram for an olivine tholeiite from San Juan Mountains, New Mexico, determined by Cohen *et al.* (1967, p. 496, Figure 4). (With permission of the *American Journal of Science*.)

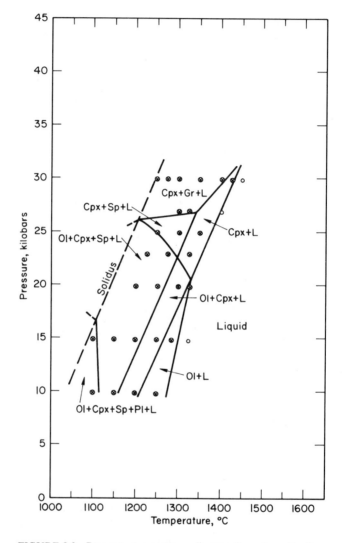

FIGURE 8-3   Pressure–temperature diagram for a basanite from Mt. Shadwell, Victoria, Australia, determined by Arculus (1975, p. 514, Figure 74A). (With permission of the Carnegie Institution of Washington.)

3. A reaction relationship in the melt, wherein olivine is consumed, occurs at high pressures.

MELTING RELATIONSHIP OF BASALT TO GARNET PERIDOTITE

If basalt is indeed the partial melting product of garnet peridotite, then the liquidus of basalt, after conversion to eclogite, should be close to the solidus of the parental material. Liquidus results for an array of basalts are compared with solidus results for an array of peridotites in Figure 8-4. Some of the peridotites would be appropriate parental material for some of the basalts; others, however, are 50–100° apart. The discrepancies, aside from experimental difficulties, may indicate that:

1. The basaltic liquid fractionated after separation from its source, and the liquidus temperature is now lower.
2. Certain compositional characteristics, such as the iron–magnesium ratios, are not necessarily appropriate for an equilibrium relationship.
3. Volatiles in appropriate concentrations may have been present in either the peridotite or the basaltic liquid, or both, and may have escaped.
4. The natural materials are sufficiently altered to cause spurious results.

MELTING RELATIONSHIP OF ECLOGITE TO GARNET PERIDOTITE

Because bimineralic eclogites form one binary join in quaternary garnet peridotite, another test may be devised to ascertain the relationship of eclogitic, hence basaltic, magmas to garnet peridotite. If the garnet peridotite system is eutecticlike, then eclogite should begin to melt at a higher temperature than garnet peridotite. On the other hand, if olivine or orthopyroxene, or both, exhibit a reaction relation with liquid, then eclogite should begin to melt at a lower temperature than garnet peridotite. It is necessary, therefore, to examine the melting relations of eclogite in detail.

MELTING OF ECLOGITE

The compositions of natural eclogites are confined to a narrow band in the Af–C–Fm diagram (Figure 8-5). These data are interpreted to

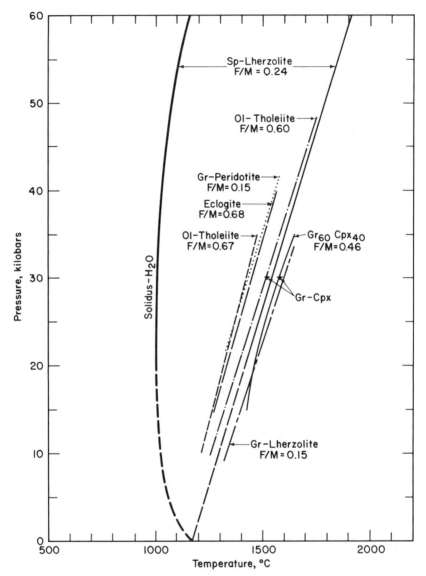

FIGURE 8-4   Comparison of the liquidus of eclogites and basalts converted to eclogites with the solidus of possible parental materials (garnet peridotite, garnet lherzolite, and spinel lherzolite). $F/M = FeO/(FeO + MgO)$

Sp-lherzolite: Kushiro *et al.* (1968a) $Gr_{60}Cpx_{40}$: Ito and Kennedy (1974)
Olivine tholeiite: Cohen *et al.* (1967)   Olivine tholeiite: Thompson (1972)
Gr-peridotite: Ito and Kennedy (1967)   Gr-Cpx: O'Hara and Yoder (1967)
Eclogite: Yoder and Tilley (1962)   Gr-lherzolite: Kushiro (1973b).

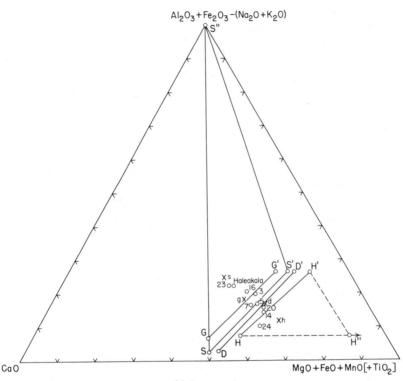

Mol percent

FIGURE 8-5 Projection of analyzed eclogites and their constituent minerals on the Af–C–Fm diagram (after Yoder and Tilley, 1962, p. 476, Figure 36).

G, G', g: clinopyroxene, garnet, and rock of Glenelg, Scotland, eclogite.

D, D', d: clinopyroxene, garnet, and rock of Loch Duich, Scotland, eclogite.

S, S', S'', s: clinopyroxene, garnet, kyanite, and rock of Silberbach, Germany, kyanite eclogite.

H, H', H'', h: clinopyroxene, garnet, orthopyroxene, and rock of Salt Lake crater, Oahu, hypersthene eclogite. (The arrow from H'' indicates composition adjusted for admixed clinopyroxene.)

X = composition of rock for which constituent minerals were analyzed.

Circle with number = composition of basalt analyzed by Yoder and Tilley (1962, p. 361–362, Table 2).

Circle labeled Haleakala = composition of basalt from Yoder and Tilley (1962, p. 417, Table 23, No. 4).

(With permission of Oxford University Press.)

define the boundary curve between the two major phases in the rock, as illustrated in Figure 8-6. Their close proximity to such a boundary curve suggests that eclogites are themselves derivative products of a more primitive material, presumably garnet peridotite. The principal question then becomes: How are the two major phases of garnet peridotite, olivine and orthopyroxene, eliminated so that its partial melt crystallizes only garnet and clinopyroxene (i.e., eclogite)?

ORTHOPYROXENE REACTION RELATION

A study of the diopside–pyrope system at 30 kbar by O'Hara and Yoder (1967, p. 74, Figure 3) provided an explanation for the elimination of orthopyroxene (Figure 5-5). At about 1620°C, orthopyroxene reacts with liquid to form garnet and clinopyroxene. The relationship is better displayed in isothermal sections at nearby temperatures (Figure 8-7). The switching of the tie lines from orthopyroxene and liquid at 1630°C (Figure 8-7A) to those connecting coexisting clinopyroxene and garnet at 1615°C (Figure 8-7B) demonstrates the reaction relationship. Experiments using natural minerals yielded the same results. In addition, the reaction relationship persists to 40 kbar, according to Davis (1964, p. 167, Figure 60). These results indicate that the separation of volatile-free magmas of basaltic composition from garnet peridotite would have to take place at depths where the pressure is less than about 50 kbar. At and above that pressure, the garnet peridotite system appears to be clearly eutectic, and olivine would crystallize out of separated liquids.

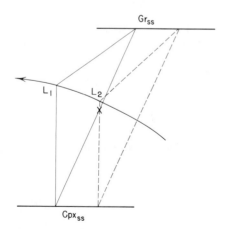

FIGURE 8-6   Schematic interpretation of melting of garnet and clinopyroxene solid solutions illustrated in Figure 8-5. X = bulk composition. L_1 and L_2 are projections of the liquid compositions at temperatures T_1 and T_2, respectively, where $T_1 < T_2$. Pressure constant. From Yoder and Tilley (1962, p. 505, Figure 47). (With permission of Oxford University Press.)

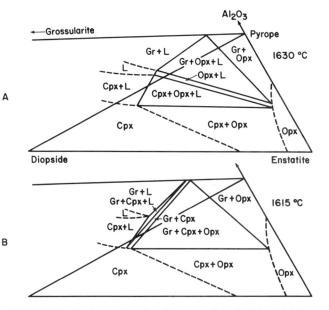

FIGURE 8-7 Isothermal sections for a part of the CaSiO$_3$–MgSiO$_3$–Al$_2$O$_3$ plane at 30 kbar. *A*, 1630°C. *B*, 1615°C. Switching of the tie lines for orthopyroxene + liquid at 1630°C with the tie lines for clinopyroxene + garnet at 1615°C indicates that the reaction orthopyroxene + liquid = clinopyroxene + garnet has taken place. From O'Hara and Yoder (1967, p. 75, Figure 4). (With permission of Oliver and Boyd.)

OLIVINE REACTION RELATION

The search for the olivine reaction relationship in garnet peridotite assemblages was unsuccessful for several years until a study by Kushiro (1968, p. 629, Figure 8) of a composition on the join enstatite–Ca-Tschermak's molecule, pursued for different purposes, exhibited an olivine reaction relationship in the pressure range 17–26 kbar. Because of the importance of this reaction to the formation of olivine-free eclogite, a wider range of compositions was investigated by Kushiro and Yoder (1974, p. 266–267). The change of the liquidus surface of a portion of the MgSiO$_3$–CaSiO$_3$–Al$_2$O$_3$ system with pressure was determined (Figure 8-8). The liquidus field of olivine at 20 kbar in the chosen plane, its presence clearly establishing the reaction relation, diminishes in size with increasing pressure. At 30 kbar only the orthopyroxene reaction relationship is exhibited. Furthermore, Davis and Schairer (1965, p. 124, Figure 35) showed that the system

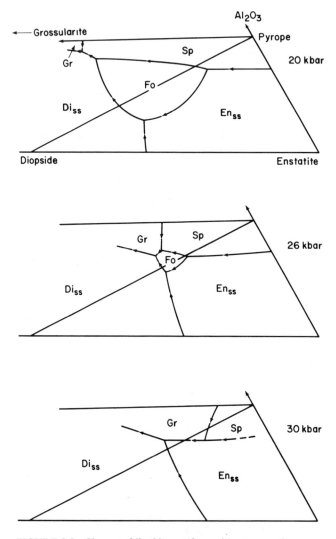

FIGURE 8-8 Change of liquidus surface with pressure for a portion of the CaSiO$_3$–MgSiO$_3$–Al$_2$O$_3$ system. Presence of forsterite on liquidus at P = 20 and 26 kbar implies a reaction relationship of forsterite with a SiO$_2$-rich liquid. Field of pigeonite deleted for simplicity. Data from O'Hara and Yoder (1967), Kushiro (1968), and Kushiro and Yoder (1974, p. 268, Figure 39). (With permission of the Carnegie Institution of Washington.)

forsterite–diopside–pyrope was close to eutecticlike at 40 kbar. The rarity of olivine-bearing eclogite nodules whether or not genetically related, supports the view that few magmas are derived from depths as great as 130 km where garnet peridotite becomes eutectic and olivine may precipitate from separated liquids.

A possible origin of eclogite from garnet peridotite is elucidated by these experiments. If allowances are made for the differences in composition between natural magmas and the simplified systems studied, magmas formed by the partial melting of garnet peridotite and separated at depths less than about 90 km (= 28 kbar) will exhibit both the olivine and the orthopyroxene reaction relationships. The liquids so produced may yield bimineralic eclogite cumulates or continue to shallower depths and yield olivine-bearing basalts. Furthermore, eclogites would melt at lower temperatures than garnet peridotite at pressures less than 28 kbar and at higher temperatures above that pressure, where a eutecticlike relationship exists in the garnet peridotite. O'Hara and Yoder (1967, p. 101), for example, observed that at 30 kbar, eclogite melts at a higher temperature than garnet peridotite and concluded that the garnet–clinopyroxene assemblage constitutes a thermal divide between forsterite-bearing assemblages and quartz-bearing assemblages. The sequence of reactions remains to be shown; that is, does orthopyroxene or olivine react out first with decreasing pressure? If olivine reacts out first in the CaO–MgO–Al_2O_3–SiO_2 system, then the liquids will become enriched in silica directly, the composition of the liquid passing to the SiO_2-rich side of the plane Cpx–Opx–Gr (see Figure 2-11). If orthopyroxene reacts out first, the initial trend is toward larnite-normative liquids, the composition of the liquid passing to the larnite-rich side of Cpx–Ol–Gr.

BASALT COMPOSITION CHANGES DURING RISE

The failure of olivine basalts to retain olivine when converted to eclogite appears to confirm an olivine reaction relation. Another explanation, however, may account for the absence of olivine. Suppose a liquid of basaltic composition is derived from a parent composed predominantly of olivine; then olivine should crystallize from the separated liquid. As the liquid rises to the surface, the proportions and composition of the precipitating phases may be expected to change in response to the drop in pressure and temperature. On the basis of the results on simple systems (discussed immediately below), olivine is the principal phase to precipitate along with other phases if the magma is to arrive at the surface with a eutecticlike composition and contain phenocrysts. That is to say, the magma has no superheat. The magma that appears at the surface is therefore not the composition originally in equilibrium with the parent but is a derivative. This point has been stressed by O'Hara (1965; 1968a, p. 686) and is his principal objection to considering tholeiitic rocks as representative of primary magmas. On these grounds it is necessary to restore the crystals lost on the way to the surface before the relation of basalt to its presumed parent, garnet peridotite, can be ascertained.

In this light, the failure to observe olivine in the high-pressure assemblage experimentally obtained from basalts crystallized at the surface can now be examined with the aid of two simple diagrams used by Yoder and Tilley (1962, p. 500, Figures 44a and 44c) to illustrate the variation of fractionation courses at high and low pressure. In Figure 8-9A are given the system $Ne-Fo-SiO_2$ and the joins at 1 atm and 30 kbar. If a parental liquid X were fractionated at successively lower pressures, the final composition would probably be in a position such as Y. Because the forsterite field expands continuously with decreasing pressure, only forsterite crystallizes from the liquid. The parental assemblage would be Fo + En + Jd, and the derivative assemblage at low pressure would be Fo + En + Ab. The recrystallization of Y at 30 kbar would yield the assemblage En + Jd + Qz; that is, no Fo would be observed. The system $Fo-CaTs-SiO_2$, shown in Figure 8-9B, can be treated in a similar way. The parental assemblage X gives way at low pressures to Fo + En + An after fractionating to Y. Recrystallization of Y at high pressures results in an assemblage devoid of Fo. The loss of olivine in basalt at high pressure could therefore be explained by the predominance of plagioclase over olivine + hypersthene, the olivine being consumed at high pressures initially through the reactions of An + Fo and An + En, investigated by Kushiro and Yoder (1966), and of Ab + Fo, investigated by Yoder and Tilley (1962).

OLIVINE—INCOMPATIBLE OR COMPLEMENTARY?

It is easily seen that the addition of Fo to the composition Y in both cases presented in Figure 8-9 must be done with considerable caution. It must first be demonstrated that Y was indeed fractionated from a Fo-bearing assemblage. The pyrolite model of Ringwood (1962 and subsequent versions) was devised by adding approximately three or four parts dunite to one part basalt on the assumptions that the dunite was the only residuum and that the basalt was the complementary primary magma from the partial melting of an olivine-bearing parent (i.e., pyrolite). The derivative nature of basalts (and eclogites) has

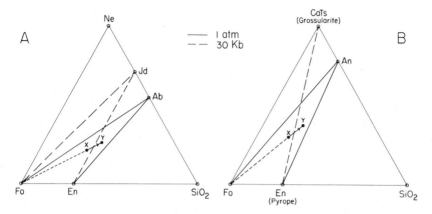

FIGURE 8-9   *A*. The joins in the Forsterite (Fo)–nepheline (Ne)–silica (SiO₂) system at 1 atm and 30 kbar (after Yoder and Tilley, 1962). Point *X* represents schematically the composition of a critical liquid formed at high pressure, and *Y* represents its derivative liquid at low pressure, assuming removal of olivine only. The small-dash line is a construction line for determining the course of liquid by removal of olivine.

  B. The joins in the forsterite(Fo)–calcium-Tschermak's-molecule(CaTs)–silica(SiO₂) system at 1 atm and 30 kbar after Yoder and Tilley (1962). Liquid compositions *X* and *Y* bear the same relationship as in *A*. End members of the garnet join (long-dash line) grossularite–pyrope project from Al_2O_3 on CaTs and En, respectively, in the CaO–MgO–Al_2O_3–SiO₂ system. From Yoder (1974, p. 265, Figure 37*A* and 37*B*). (With permission of the Carnegie Institution of Washington.)

already been demonstrated by Yoder and Tilley (1962) and emphasized by O'Hara (1968b). Thus, the addition of olivine to a derivative rock with which it may not be compatible at high pressures does not appear to be a suitable model for deriving primary magmas.*

EVIDENCE FROM NODULES

That olivine and orthopyroxene are the principal phases in nodules brought up by some *alkali* basalts supports the belief that these minerals are important phases in the parental material of basalts. In addition, Huckenholz (1965, 1966, 1973) observed relic orthopyroxene in the nepheline-bearing olivine basalts and basanites of the Hocheifel area, Germany. The preservation of the nodules and relics suggests that

---

*These arguments do not conflict with a model based on the addition of olivine to an olivine-bearing, quartz-normative tholeiite derived by the separation of magma from a plagioclase peridotite at low pressures (= depths < 15 km) where the olivine reaction relation is operative.

such magmas were brought directly to the surface from the mantle source region. The nodules contain at least one phase, orthopyroxene, with which the host rock is incompatible at low pressures. In high-pressure experiments Kushiro (1964, p. 110, Figure 27B) found that enstatite and nepheline can coexist; therefore, a nodule with orthopyroxene could be compatible with a nepheline-normative liquid at certain depths.* The magma was effectively separated from most of the parental material at high pressures, at temperatures below both the olivine and orthopyroxene reaction points, or at the latter reaction point.

The olivine-free eclogite nodules are probably derived as cumulates in the 60–85-km region where the required reaction relationships exist under anhydrous conditions, according to the experiments of Kushiro (1968). Alternatively, some eclogites could be recycled derivative basalts. For example, the metamorphosed oceanic basalts in a subducted plate might be an adequate source for those eclogites.

FIRST PHASE CONSUMED IN MELT

With the data now presented, it is appropriate to return to the question of which of the two least abundant major phases in garnet peridotite is consumed first in a partial melt. It was observed in Figures 8-2, 8-3, and 8-4 that clinopyroxene was on the liquidus of basaltic rocks at intermediate pressures and garnet was on the liquidus at the highest pressures investigated. These observations are supported by the shift of the liquidus boundary between 30 and 40 kbar for the diopside–pyrope system (Figure 8-10): the boundary curve (Figure 8-6) shifts toward clinopyroxene with increasing pressure. For a bulk composition with clinopyroxene and garnet in the proportions $Cpx_{50}Gr_{50}-$

---

*The relationship was believed to have been confirmed in high-pressure experiments on synthetic compositions simulating alkali basalt (Green and Ringwood, 1967), picritic basanite, and picritic nephelinite (Bultitude and Green, 1971), respectively, having normative nepheline equal to 2.2, 9.4, and 10.6 wt%. However, a natural basanite (Ne = 10.8 wt%) studied by Ito and Kennedy (1968, p. 186, Figure 3), two natural basanites (Ne = 7.8 and 11.9 wt%) investigated by Arculus (1975), and a synthetic composition simulating olivine nephelinite (Ne = 15.3 wt%; Bultitude and Green, 1971) did not yield orthopyroxene at high pressures. The effects of bulk composition, H_2O and CO_2 from parts in the high-pressure apparatus, iron loss to the sample container, oxidation, and metastability may account for these differences. The appearance of orthopyroxene in nepheline-normative compositions at high pressures was erroneously attributed to the presence of H_2O by Bultitude and Green (1971), contrary to the results on simple hydrous systems. Eggler (1973) demonstrated that the effect was due primarily to the presence of CO_2. Experiments on a series of natural nepheline-normative rocks under more carefully controlled conditions are warranted.

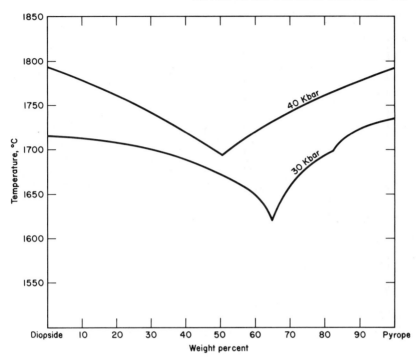

FIGURE 8-10   The liquidus of the diopside–pyrope system at 30 kbar (after O'Hara and Yoder, 1967, p. 74, Figure 3), ignoring a field of enstatite, and at 40 kbar (after Davis 1964, p. 167, Figure 60). The liquidus phase at and near pyrope at 30 kbar is spinel. The shift of the coprecipitation curve for garnet and clinopyroxene with pressure is illustrated. (With permission of Oliver and Boyd and the Carnegie Institution of Washington.)

$Cpx_{35}Gr_{65}$, it is apparent that at the lowest pressures where garnet peridotite forms a stable assemblage, garnet is consumed first, and with increasing pressure clinopyroxene is consumed first, with partial melting. These shifts in the boundary curve have considerable influence on the normative character of the liquid. Some of the principal normative minerals in an analyzed clinopyroxene and garnet from a garnet peridotite are plotted in Figure 8-11. The shift in normative character of a potential melt resulting from changing proportions of clinopyroxene and garnet in the liquid is evident. With increasing garnet contribution to the melt, for example, the normative hypersthene and anorthite content increases, whereas the normative diopside content decreases within a reasonable range such as $Cpx_{70}Gr_{30}$–$Cpx_{30}Gr_{70}$ (wt%).

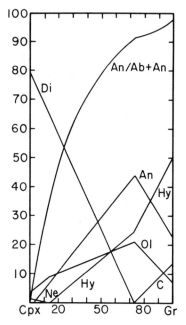

FIGURE 8-11 Norms calculated for mixtures of analyzed coexisting clinopyroxene and garnet from a garnet peridotite from South Africa (O'Hara and Yoder, 1967, p. 104, Figure 9). C = corundum.

RARE-EARTH ELEMENT DATA

MEASURE OF DEGREE OF PARTIAL MELTING?

The rare-earth element analyses normalized to the average of 20 chondrites (see Appendix) for three basalt types in Hawaii are plotted in Figure 8-12. Similar rare-earth element patterns were obtained by Arculus and Shimizu (1974, p. 558, Figure 211) for a suite of basanitoids and alkali olivine basalts ranging in normative composition from 8.99 percent Ne to 4.80 percent Hyp. These curves were interpreted by Schilling (1971) to be indicative of the degree of melting, the top curve representing a small amount of partial melting and the bottom curve, a more advanced stage of partial melting. To accommodate the major element oxide compositions and thermal divides, however, Schilling (1966, p. 79) also indicated that the magmas would have had to form at different depths or pressures. These conclusions must be tempered with the assumption that the composition of the liquid from which the rocks crystallized did not change en route to the surface. An examination of the normative character of these three rocks does not support the view that they were produced by varying degrees of partial melting from a single parent at the same pressure. The critical normative contents in weight percent are:

Rock	Ol	Hyp	Ne	La
Melilite nephelinite	18.35	—	21.58	7.40
Ankaramite	23.88	—	3.12	—
Olivine tholeiite	6.75	28.84	—	—

Normative Component

If the melilite nephelinite is the product of a very small amount of melting, then it is not likely that larger amounts of melting will yield an olivine tholeiite with such a small amount of olivine unless the pressure is greatly reduced or H_2O plays a significant role. Both H_2O and reduced pressure tend to enlarge the field of olivine, thereby reducing the olivine content of liquids on its boundary curves and surfaces. On the other hand, the ankaramite, assuming it is not a cumulate, might be the product of a greater amount of melting at the same pressure of the same parent that produced the melilite nephelinite, because of the increase in olivine content. Increased melting at the same pressure of a rock in which olivine is the most abundant phase would progressively lead to an increase in the olivine content of the liquid. It appears that these rocks cannot be related simply to a parental material solely as products of different degrees of melting at the same pressure. The series of rocks studied by Arculus and Shimizu (1974) are much more restricted in normative character and exhibit a small but progressive increase in olivine content from 20.45 percent to 24.38 percent with moderate deviations.

A theoretical analysis by Schilling and Winchester (1967) of the sequence of rare-earth element changes with degree of partial melting yields the patterns shown in Figure 8-13. Although these calculated curves have not been correlated with the normative character of the deduced liquids, the curve for the relative rare-earth composition of liquid at 1 percent melting, $L(1\%)$, would approximately express the character of the data obtained for melilite nephelinites. The curve $L(50\%)$ is similar to that obtained for Hawaiian tholeiites. In Schilling's view the rare-earth element data constitute the strongest argument that the degree of melting influences the normative character of the melt.

Green (1971, p. 709–710, Figures 1 and 2) apparently accepted this view and proposed that within the pressure range 15–35 kbar the normative character changes from larnite to nepheline to hypersthene merely by increasing the proportion of melt at constant pressure with

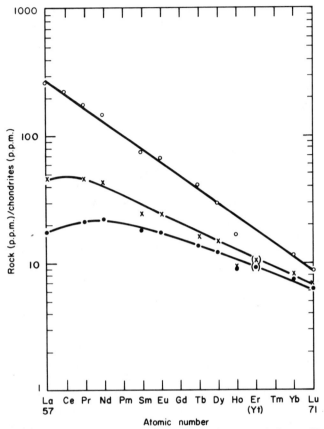

FIGURE 8-12   Abundances of rare-earth elements relative to 20 chondrites in a melilite nephelinite (open circles: normative La = 7.40, Ne = 21.58), ankaramite (X: normative Ne = 3.12), and olivine tholeiite (black dots: normative Hyp = 28.84) from the Hawaiian Islands plotted on a logarithmic scale as a function of atomic number. From Schilling and Winchester (1967, p. 269, Figure 2). (With permission of John Wiley and Sons, Inc.)

as little as 0.01 percent H_2O! In all the experimental studies of the simplest to the most complex synthetic systems, little phase-equilibrium evidence has been found to support this view. No single bulk composition has yet been shown to produce a complete series of La–Ne, Ol–Ne, Ol–Hyp, and Hyp–Qz normative liquids merely by increasing the degree of melting over a moderate range (e.g., 30 percent) or, in fact, over any range of melting. It was demonstrated with the forsterite–diopside–pyrope system that the degree of melting does influence the composition of liquid, but within narrow normative bounds. The proportion of volatiles and the pressure, on the other hand, greatly influence the normative character of the melt (Mysen, 1975). Yet the presence of a gas phase does not appear to greatly influence the partitioning of the rare-earth elements (Cullers *et al.,* 1970), nor does the accompanying shift of "eutectic" composition (Schilling, 1975, p. 1467, Figure 9).

It seems fortuitous that the required degree of melting could be achieved each time to produce successive batches of the same compo-

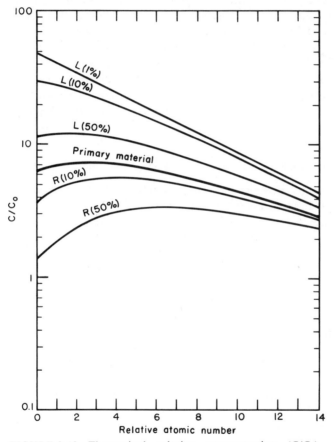

FIGURE 8-13 Theoretical relative concentration (C/C_0) (logarithmic scale) of the 14 rare-earth elements (0 = lanthanum; 14 = lutecium) in melts (L) and residua (R) resulting from the partial melting of primary material having the illustrated composition for 1 percent, 10 percent, and 50 percent melting. From Schilling and Winchester (1967, p. 277, Figure 9). (With permission of John Wiley and Sons, Inc.)

sition. The only potential control of degree of melting identified is the ill-defined proportion of crystals to liquid that leads to the physical disaggregation of the parental material. Schilling (1966, p. 80) believed that the degree of melting could be determined by the abrupt change in heat requirements attending the loss of a phase on *partial* melting. The sharp rise in liquidus temperature, particularly that of olivine, may place a temporary limit on the degree of melting. On *fractional* melting, the loss of a major phase, however, greatly changes the subsequent liquid composition, and a basaltic liquid may not be a likely product. Furthermore, the occurrence within the Nazca Plate of nepheline-normative basalts and olivine tholeiites with the *same* rare-earth element pattern (Schilling and Bonatti, 1975) casts doubt on the concept that the degree of melting is the principal control on the composition of liquid produced. For that region, Schilling and Bonatti (1975) found it necessary to assume that the two magma types originated at different eutectic compositions but as products of approximately the same degree of melting.

MEASURE OF HETEROGENEITY OR PROPORTION OF PHASES?

What, then, does produce the systematic variation in the patterns of abundance of rare-earth elements for the major basaltic rock types if the degree of partial melting is not always the important factor? One factor is the abundance of the rare earths in the parental material. A heterogeneous mantle would satisfy the variance, but would the variance be systematically correlated with the many other factors that determine magma type? A more appropriate dependence appears to be on the proportion of phases taking part in the partial melting. Representative distribution coefficients for the rare-earth elements between crystal and liquid are given in Figure 8-14. The representative partition coefficients for the rare-earth elements between "crystal" and "liquid"* for the major phases in garnet peridotite were summarized by Helmke and Haskin (1973, p. 1520, Figure 3). The partition coefficients between synthetic clinopyroxene and liquid as well as synthetic garnet and liquid were determined, respectively, by Masuda and Kushiro (1970, p. 44, Table 2) and Shimizu and Kushiro (1975, p. 413, Table 1). It is evident that the light rare-earth elements will always be enhanced in the liquid relative to the residuum. On the other hand, the heavy rare-earth elements may be depleted or enhanced in the liquid, depending on whether garnet persists as a phase during the melting process.

*The "crystals" are phenocrysts from natural rocks that are presumed to be in equilibrium with "liquid" represented by the groundmass or calculated residual liquids.

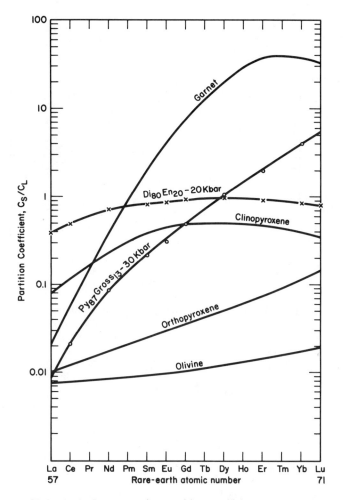

FIGURE 8-14 Representative partition coefficients for the rare-earth elements between natural crystals (olivine, orthopyroxene, clinopyroxene, and garnet) and deduced liquid compiled by Helmke and Haskin (1973, p. 1520, Figure 3), between synthetic clinopyroxene ($Di_{80}En_{20}$) and glass at 20 kbar (Masuda and Kushiro, 1970, p. 44, Table 2), and between synthetic garnet ($Py_{87}Gross_{13}$) and glass at 30 kbar (Shimizu and Kushiro, 1975, p. 413, Table 1). C_S = concentration of rare-earth element in the residual crystal; C_L = concentration of rare-earth element in coexisting liquid.

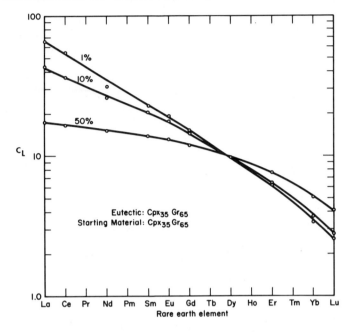

FIGURE 8-15 Calculated rare-earth element concentrations in liquid, C_L, generated by 1 percent, 10 percent, or 50 percent partial melting of a hypothetical eclogite eutectic composition ($Cpx_{35}Gr_{65}$) having 10 times the normalized average chondrite abundances of rare earths (see Appendix). The partition coefficients for synthetic clinopyroxene–liquid and garnet–liquid given in Figure 8-14 were used, and pressure difference was neglected.

It is evident from Figure 8-14 that major changes in the rare-earth element pattern will result from variations in the proportion of clinopyroxene and garnet. Variation in the content of olivine, the presumed dominant phase, will have little influence on the form of the rare-earth element patterns, and yet it will greatly affect the overall abundance of rare-earth elements in the liquid. Consider, then, a simple binary system, clinopyroxene–garnet, with a eutectic at $Cpx_{35}Gr_{65}$ (*cf.* Figure 5-5). The experimental partition coefficients given in Figure 8-14, although they were determined at different pressures and may not be compatible, can be used to calculate the rare-earth element contents of liquids according to the relationship given by Schilling (1975, Appendix I):

$$C_L^i = \frac{C_P^i}{Y \left(1 - \sum_{j=1}^{j=n} K_j^i E_j\right) + \sum_{j=1}^{j=n} K_j^i X_j},$$

where

i = rare-earth element
j = phase
C_L = concentration of rare-earth element in liquid
C_P = concentration of rare-earth element in parental material
Y = weight fraction of liquid (i.e. degree of melting)
K = partition coefficient between residual phase and liquid
E = eutectic proportion in weight percent of phase j
X = fraction of residual phase remaining in equilibrium with liquid in weight percent

The rare-earth element pattern of the liquid appears to depend on not only the degree of melting, but also the initial abundances, the proportion of phases in the parental material, and the proportion of phases contributing to the liquid. If the eclogite (Cpx + Gr) undergoing melting has an initial rare-earth element abundance 10 times that of the average chondrite, then the rare-earth element patterns calculated for various degrees of partial melting of the eutectic composition are as presented in Figure 8-15. If a constant degree of melting of 1 percent is assumed, then the influence of various initial proportions of phases is as displayed in the calculated rare-earth patterns of Figure 8-16.* It is clear that changing the proportion of critical phases is as important as the degree of melting (Schilling, 1975, p. 1466). Gast (1968, p. 1071) was of the opinion that the relative concentrations of rare-earth elements were "quite insensitive to the phase that melts." Aside from the fact that all phases must participate in the melt, the conclusion appears to be in error because it was based on calculated changes dominated by olivine and orthopyroxene content. Even though the more abundant phases in the parental material have the greatest influence on the rare-earth composition of the melt, the changes in character of the

---

*At first glance, it may seem unreasonable to assign the same initial rare-earth abundances to differing proportions of the same minerals. Many eclogites do appear to have chondriticlike abundances, and one need only look at the chondrites themselves for examples of material having the same rare-earth abundances yet varying greatly in the proportions of phases.

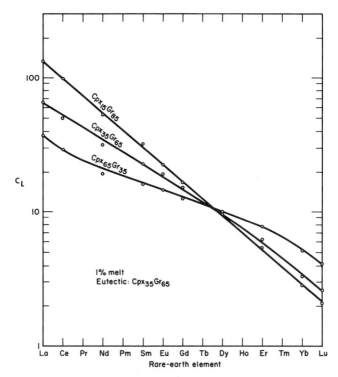

FIGURE 8-16   Calculated rare-earth element concentration in liquid, C_L, generated with 1 percent partial melting of different proportions of clinopyroxene and garnet in a hypothetical eclogite having 10 times the normalized average chondrite abundances of rare earths (see Appendix). The eutectic composition of the system was assumed to be $Cpx_{35}Gr_{65}$. The partition coefficients for synthetic clinopyroxene–liquid and garnet–liquid given in Figure 8-14 were used, and pressure difference was neglected.

rare-earth patterns of the melt reside primarily in the contributions from clinopyroxene and garnet (Figure 8-14). There is a clear need to determine the effect of the variables experimentally and evaluate their influence on the rare-earth element pattern. Nevertheless, the rare-earth elements appear to record information not yet identified among the major element relations.

RARE EARTHS IN GARNET PERIDOTITE

The distribution of rare earths in a hypothetical garnet peridotite and its melts and residues has been computed by Schilling (1975, p. 1469,

Figure 11*b*). He assumed that the garnet peridotite consisted of $Ol_{57}Opx_{17}Cpx_{12}Gr_{14}$ and that the eutectic composition was Ol_3Opx_3-$Cpx_{47}Gr_{47}$ (after Davis and Schairer, 1965). One model yielded the results shown in Figure 8-17 for 10 percent and about 30 percent* partial melting. The principal observation is that the heavy rare earths are retained in the residuum after 10 percent partial melting, whereas the light rare earths are depleted after the same amount of partial melting. This model demonstrates that garnet is present initially, persists with small amounts of partial melting, and is consumed with somewhat larger amounts of partial melting. Schilling pointed out that no natural ultramafic rock has yet been found that exhibits the implied enrichment of heavy rare earths representative of residuum from very small amounts of partial melting of a garnet-bearing parent.

GENERAL REMARKS AND SUMMARY

In conclusion, the absence of olivine in basalts held at the high pressures at which they are believed to be derived from garnet peridotite may be due, on the one hand, to the derivative nature of the basaltic magmas. The primary magmas, partially melted and separated from garnet peridotite, change composition en route to the surface and result in a liquid composition that, when recrystallized at high pressure, does not contain olivine. On the other hand, eclogite, equivalent in composition to basalt, requires both an olivine and an orthopyroxene reaction relationship with liquid if it is derived from garnet peridotite at high pressures. The orthopyroxene reaction relationship was demonstrated by O'Hara and Yoder (1967), and the olivine reaction relationship was demonstrated by Kushiro and Yoder (1974). An eclogite liquid, brought to the surface and crystallized as basalt, would therefore also be derivative.

The few data available on the liquidus phase of basaltic compositions at high pressure suggest that clinopyroxene is consumed first in partial melting of garnet peridotite at high pressures, whereas garnet is consumed first at relatively moderate pressures. These conclusions are based on the assumption that H_2O does not play a significant role in basaltic magma generation.

The rare-earth-element data may yield information on the degree of melting, the proportions of phases in the starting material, the presence or absence of garnet, and the extent of heterogeneity in the upper mantle. Experimental evaluation of the principal variables is needed

*Clinopyroxene is consumed after 25.5 percent melting (Schilling, personal communication, 1976).

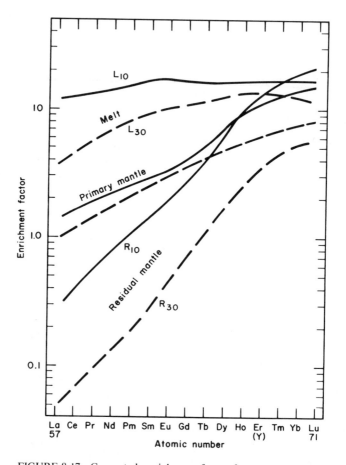

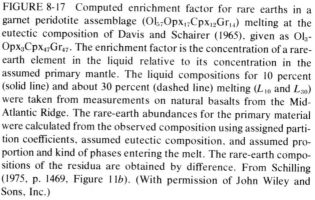

FIGURE 8-17 Computed enrichment factor for rare earths in a garnet peridotite assemblage ($Ol_{57}Opx_{17}Cpx_{12}Gr_{14}$) melting at the eutectic composition of Davis and Schairer (1965), given as Ol_3-$Opx_3Cpx_{47}Gr_{47}$. The enrichment factor is the concentration of a rare-earth element in the liquid relative to its concentration in the assumed primary mantle. The liquid compositions for 10 percent (solid line) and about 30 percent (dashed line) melting (L_{10} and L_{30}) were taken from measurements on natural basalts from the Mid-Atlantic Ridge. The rare-earth abundances for the primary material were calculated from the observed composition using assigned partition coefficients, assumed eutectic composition, and assumed proportion and kind of phases entering the melt. The rare-earth compositions of the residua are obtained by difference. From Schilling (1975, p. 1469, Figure 11b). (With permission of John Wiley and Sons, Inc.)

before unambiguous conclusions can be reached. In general, trace elements indicate that the peridotites in hand are depleted in varying degrees in the constituents of basalt. Furthermore, some basalts, especially the oceanic basalts, appear to be the product of secondary or tertiary melting events or the later product of fractional melting. Other basalts, the alkali types, appear to be the product of primary melting at an appropriate invariant point or an early product of fractional melting.

9 Tectonophysics of Melting

LOCATION OF INITIAL LIQUID

Whatever the parental material was prior to the beginning of melting, it went through one or more metamorphic reactions as the temperature was raised. Garnet peridotite can be expected to have the granular texture of a metamorphic rock (Plate 2, p. 44b). The interfacial angles between the grains are approximately 120° on the average, according to Vernon (1968; 1970), but vary because of the anisotropy of the crystals (Figure 9-1). "The points of intersection" of three grains in thin section mark a potential channel or vein. The junction of four grains is the possible site of a tetrahedral pore. The grain contacts can be viewed as local binary, ternary, and quaternary systems; yet the rock begins to melt at the temperature appropriate for the whole rock system.

Melting involves all phases, and there obviously must be communication between adjacent and more distant neighboring grains by diffusion processes. The very first melt may actually begin within inclusions or involve minor phases (e.g., apatite) interstitial to the major phases. Fluid inclusions tend to migrate to crystal boundaries up thermal gradients, and liquid–vapor inclusions tend to migrate down thermal gradients (Anthony and Cline, 1970, 1971, 1972a,b). As a result of the failure of the rock to reach textural and chemical equilibrium during the metamorphic stage preceding melting, the crystals may also retain chemical zoning or impurities at their periphery that would tend to

162

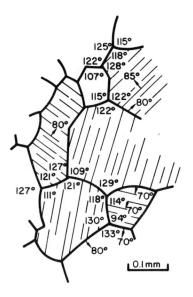

FIGURE 9-1 Measured interfacial angles and dips of cleavage (where not approximately vertical) of hedenbergite aggregate in a high-grade metamorphic rock from Broken Hill, Australia. From Vernon (1968, p. 8, Figure 6). (With permission of the Oxford University Press.)

lower the melting temperature at the grain boundary relative to the interior of the grain. It is believed that pervasive melting begins at those grain junctions where all phases are in contact. In this sense some models illustrate the intial melt as isolated droplets. With an increase in thermal energy the first liquid provides communication to other grains. If the rock is mineralogically homogeneous, all grains will be in equilibrium with that first liquid.

ACCUMULATION OF LIQUID

The interior of an initial liquid droplet, if that form ever obtains, is subjected to an additional pressure resulting from the surface tension. According to Coble and Burke (1963, p. 206), the pressure is given by

$$P = 2\gamma r^{-1},$$

where γ = surface tension and r = radius of curvature of the droplet. For ceramic materials, γ equals 1,000 erg/cm^2; therefore, a 1-μm droplet is under an additional confining pressure of 40 bars by virtue of its size and shape. The external hydrostatic pressure is thereby acting to reduce the size of the drop. The implication is that the liquid does not wet the contiguous solid surfaces well. On the other hand, if the liquid does wet the solid surfaces well, then the liquid will be drawn into the re-entrant angles between the contiguous solid surfaces along which melting is taking place (Smith, 1948, p. 20, Figure 5). The result will be a "liquid-bonded aggregate" in the terms of the ceramicists (Allison *et al.*, 1959, p. 517). All grains would then be initially coated with a thin film of liquid. The wettability of different crystals will differ and in different crystallographic directions. The physical properties of a rock undergoing initial melting will no doubt differ greatly depending on how the liquid is distributed. In the writer's view, a three-dimensional network of liquid film is the preferred description of the initial melt, in contrast to an array of isolated droplets of variable aspect ratio.

There are no experimental observations on the succeeding events in the melting of a garnet peridotite; some insight may be gained, however, by examining the results obtained on a quartz-feldspar gneiss. Mehnert *et al.* (1973) described the initial anatectic stages on the basis

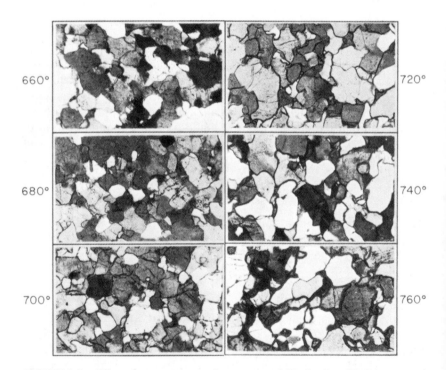

FIGURE 9-2.   Effect of temperature on the amount and distribution of melt in a natural quartz-feldspar gneiss held at $P_{H_2O} = 2$ kbar for 24 hours at each temperature. Width of section, approximately 1 mm. From Mehnert *et al.* (1973, p. 168, Figure 1). (With permission of E. Schweizerbart'sche Verlagsbuchhandlung.)

of hydrothermal experiments at a series of temperatures, pressures, and times. One set of experiments at constant water pressure (2 kbar) and time (1 day) shows the development of liquid with increasing temperature (Figure 9-2). At about 680°C very thin films of glass, about 1 μm or less in thickness, can be seen locally, particularly at the contacts of the three principal phases. At successively higher temperatures, the films increase in thickness uniformly, irrespective of the composition of the adjoining crystals. Concentration of the low-temperature melting residua and impurities along grain boundaries in the final stages of the original crystallization of the rock no doubt contributes greatly to the early development of a film of liquid around all the grains of the rock on remelting (see McLean, 1957, pp. 100–107). It appears that migration along cracks, cleavages, grain boundaries, and other discontinuities occurs readily. The initial partial melting of granite under a water pressure less than the total pressure also yields an almost continuous network of films, according to Brace *et al.* (1968).

Further support of the "liquid-bonded aggregate" concept of initial rock melting is also found in the study of simple oxide systems. A synthetic mixture in the $CaO-MgO-Al_2O_3$ system was held at a suitable temperature by Clark *et al.* (1953, pp. 32–33) so that 10 percent melt was produced. Single-crystal grains of periclase were observed to be separated by films of liquid. They particularly noted the absence "of separate 'lakes' of liquid in a continuous solid matrix."

HOMOGENEITY OF LIQUID

As melting progresses, the films unite,* and the crystals and crystal aggregates form irregular relics. Mehnert *et al.* (1973) carried out electron microprobe studies of the composition of the film and concluded

that the initial melts produced at relatively low temperatures are rather constant in composition, but heterogeneous with respect to inclusions of sub-microscopic crystalline relics. At higher temperatures, these relics vanish but the melt is heterogeneous in another sense, i.e., concentration gradients of the respective elements can be observed across the melt from one mineral contact to the other.

Such gradients may be eliminated in experiments of longer duration and under higher water pressures where mixing and flow may be effected. According to Butler (1961), the natural fusion of an arkose by a dolerite plug has produced a melt (now observed as a glass) that reveals no

*The high probability of intersection of isolated droplets and the formation of a network structure has been demonstrated by Haller (1965).

inhomogeneity of the potash content even though there is no textural evidence to indicate mechanical mixing.

ANALOGY TO GLACIAL WATER

Some appreciation for the nature of the veining of a rock undergoing melting may be gained from the presence of water in glacial ice, even though the pressure effect and volume change are of opposite sign to those for silicates. Nye and Frank (1973) emphasized the three-dimensional aspects of the channels formed at three-grain intersections in temperate ice. Deformation and recrystallization, in the view of Lliboutry (1971), tend to close off capillary intergranular channels and make glacier ice, especially firn, "practically impermeable." On the other hand, Shreve (1972) pictured deformation as aiding the passages to expand and contract. The larger channels gradually increase in size because of the heat generated by viscous dissipation and the heat transported by the flow of melt having higher temperatures. He visualized a network of passages tending in time to become "arborescent" like a river system. The main "streams" trend in the general direction of flow—upward in the case of magmas. The same picture on a large scale was presented by Wilshire and Shervais (1975, p. 270, Figure 14) in describing the anastomosing feeders to the main conduits of a volcano (Figure 9-3).

ROLE OF DEFORMATION

As in glaciers, deformation of the parental material plays a major role in the collection of magma. The disposition of the melt will be influ-

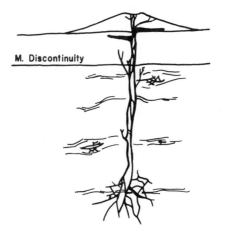

M. Discontinuity

FIGURE 9-3 Schematic diagram illustrating anastomosing feeders to main conduits in a volcanic edifice. After Wilshire and Shervais (1975, p. 270, Figure 14). (With permission of Pergamon Press, Ltd.)

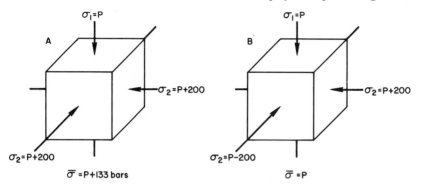

FIGURE 9-4 Comparison of average pressures acting on a unit of rock under a hydrostatic load pressure P with small (200 bars) incremental changes in horizontal stresses. Equal horizontal compressive stresses of the same sign (A) and of opposite sign (B).

enced by the shear stresses acting on the parental material, especially if flaws are present. The mean stress $\bar{\sigma}$ in a rock is equal to $1/3(\sigma_1 + \sigma_2 + \sigma_3)$, where σ is a principal stress. If the vertical stress equals P and the horizontal stresses are both $P + 200$ bars, the mean stress is $P + 133$ bars (Figure 9-4A). In the anhydrous melting of rocks the dT/dP is positive; therefore, an increase in the mean stress will cause melting to stop. There appears to be a dilemma: Any unbalanced nonhydrostatic stress tending to move magma will result in freezing the magma! One possible explanation is that the deviatoric, horizontal stresses are of opposite sign (Figure 9-4B), the rock can maintain its mean stress, and melting continues. On this basis every principal compression will be accompanied by a principal dilatation. Another possible explanation is that the strain rate is sufficient for the nonhydrostatic stress (200 bars) to generate heat, and melting ensues, provided the viscous heating exceeds the heat conduction away from the site. Still another explanation is that freezing is controlled by the rate at which the enthalpy of freezing can be conducted away. If heat conduction is slow relative to the strain rate, the magma will retain its fluidity and can move in response to the applied stress.

LOCALIZATION OF LIQUID

RESPONSE TO STRESS

Analyses by Nye (1967) and Savage (1969, p. 119) illustrate how a stress acting on a rock would cause fluid to collect in lenses parallel to

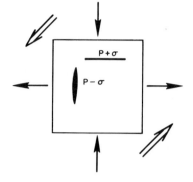

FIGURE 9-5 Response of two liquid-filled lenses perpendicular to the maximum and minimum principal stresses acting on a unit of rock under an average lithostatic pressure. Total fluid pressure decreases in lenses perpendicular to the minimum stress; and, in an anhydrous rock, melting is enhanced in those lenses. After Savage (1969, p. 119, Figure 4). (With permission of Elsevier Scientific Publishing Company.)

the principal stress (Figure 9-5). The flow of melt to regions of relatively lower pressure is appropriate because of the positive dT/dP, but what then is the flow of hydrous magma? In the presence of water, it would seem that the magma would flow into lenses perpendicular to the maximum compressive stress because higher pressure enhances melting. A detailed analysis of the hydrous case is required; present opinion, however, is that the mechanical state of a flaw or lens (open or closed) is more important than the physicochemical effect on the melting itself.

COALESCENCE

Some nodules retrieved from kimberlite pipes are extremely sheared—almost mylonitized (Figure 9-6). According to Weertman (1972), shearing is most effective in coalescing the initial droplets in a magma. On the basis of his earlier analysis of gas-bubble coalescence in a deforming glacial ice mass (Weertman, 1968), he illustrated (Figure 9-7) "bubbles" coalescing as a result of differential movement. In the light of numerous high-pressure experiments on minerals associated with garnet peridotite in solid-media apparatus, alleged to maintain hydrostatic conditions but in fact having a shear component, the initial liquid is only very rarely observed to be dispersed as droplets or "bubbles." On the other hand, application of the concept to the coalescence of "pools" of magma in an upper mantle undergoing shear over large distances and through a thick zone may be more appropriate. Shaw (1969) assigned an important role to "viscous failure and flow" in the accumulation of magma. He described (1969, p. 533) the melt fraction as being "virtually kneaded from the crystalline source." The separation of phases attributed to mechanical deformation, how-

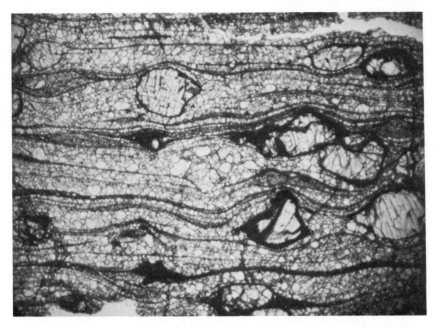

FIGURE 9-6   Sheared garnet peridotite, No. 1611. Thaba Putsoa, Lesotho. Width of section is approximately 12 mm. Photograph courtesy of Dr. F. R. Boyd.

ever, may result from other processes. The common segregation from an initial homogeneous rock of quartz and feldspar into layers alternating with layers of the remaining ferromagnesium minerals has been cited as the result of deformation during metamorphic differentiation. The observed dissimilar rheology of quartz and feldspar casts doubt on this concept (see also Shelley, 1974). Means and Williams (1974) attributed such metamorphic differentiation in a salt–mica mixture to enhanced solubility, due to stress concentrations, and redeposition.

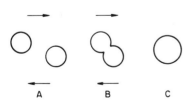

FIGURE 9-7   Coalescence of melt globule resulting from differential relative motion of individual globules. *A*, differential movement of two globules; *B*, intersection and coalescence of two globules; and *C*, spheroidization of two globules. After Weertman (1972, p. 3531, Figure 1). (With permission of the Geological Society of America.)

VOLUME CHANGE OF MELTING

LARGE INCREASE IN VOLUME

In the absence of plastic flow, shearing usually results from multiple fracturing, and fracture produces dilatancy—that is, an increase in the volume occupied by the rock. Melting also produces a large volume change. The amount of volume increase for relevant end-member minerals is given in Table 9-1. For example, the volume increase on melting the eutectic composition of the forsterite–diopside–pyrope system at 40 kbar is about 16 percent. Correction for compressibility and thermal expansion would no doubt reduce that value somewhat, but it remains clear that a large volume change must be accommodated as melting takes place. If the requisite volume is not available, the porosity being negligible, the pressure will rise and melting will stop. Although there are no data for the volume changes taking place under hydrous melting, it would appear that such melting may be enhanced if confined.

TUMESCENCE, RUPTURE, OR PLASTIC FLOW?

To accommodate the volume change on melting, a region could undergo tumescence. The inflation could lead to a rise of tens of meters at the surface, depending on the volume of melt and structural compensations. Alternatively, the volume change may be accommodated by plastic flow. If heating continues and the local stress increases further, the resultant long-range stress difference may lead to plastic flow, and

TABLE 9-1   Volume Change on Melting at 1 atm

End-Member Mineral	ρ, xtl	ρ, Glass	ΔV (cm³/gm)	% Change
Forsterite	3.223	3.035[a]	0.019[b]	6.2
Fayalite	4.068	3.764[a]	0.020[c]	8.1
Clinoenstatite	3.183	2.735	0.051	16.4
Diopside	3.275	2.846	0.046	15.1
Pyrope	3.582	3.031	0.051	18.2
Anorthite	2.765	2.700	0.010	2.7
High albite	2.605	2.382	0.036	9.4
High sanidine	2.597	2.400	0.036	8.2

[a]Calculated from estimated ΔV and ρ_{xtl}.
[b]Estimated using measured $dT/dP = 4.77°$/kbar (Davis and England, 1964, p. 1114), melting temperature = 1890°C, ρ_{xtl}, and $\Delta H_m = 208.2$ cal/g.
[c]Estimated using measured $dT/dP = 6.5°$/kbar (Lindsley, 1967, p. 230), melting temperature = 1205°C, ρ_{xtl}, and $\Delta H_m = 108.1$ cal/g.

the requisite volume will be provided by adjacent porous regions or by means of tectonic movements. The pressure developed when melting takes place at constant volume appears, at first glance, to be considerable. For diopside the volume change on melting is 0.046 cm^3/g, and the specific volume is 0.305 cm^3/g. Using the measured crystal compressibility $\beta = (1/V_0)(\Delta V/\Delta P)$, which is 1.07×10^{-6}/bar for diopside (Adams and Williamson, 1923), the potential pressure developed on melting is 140 kbar! As shown below, such circumstances never obtain.

STATE OF STRESS IN EARTH

Rocks obviously cannot withstand such high local pressures, which would cause a large deviatoric stress. Bullen (1947) estimated, on the basis of the energy release in large earthquakes, that the maximum stress in rocks did not exceed 1 kbar. Kaula (1963) said that the form of the geoid measured from satellite data indicated the earth could sustain stress differences of only 0.2 kbar. Birch (1964) noted that the stress differences associated with gravity anomalies of 200 mGal were about 0.5 kbar at depths of 50 to 100 km. Brace (1964) measured the tensile strength of diabase in the laboratory and found it to be only 0.4 kbar, whereas Roberts (1970, p. 297) thought that the tensile strength of the upper mantle was about 0.5 kbar and that the stress difference for shear fracture would therefore be 4 kbar. Jeffreys (in Runcorn, 1967, p. 433–435) argued for a strength of 1 kbar at the surface and 0.1 kbar at 600 km. [The deepest earthquakes are recorded at 700 km (Griggs and Handin, 1960, p. 359).]

It is evident that the pressures developed on melting will in fact deform or break the confining rock, but will the melting take place in view of the positive dT/dP? Under constant volume conditions, the temperature–pressure gradient is defined by

$$\left.\frac{dT}{dP}\right|_V = \frac{\beta}{\alpha},$$

where β = compressibility and α = thermal expansion. For diopside the $dT/dP|_V = 45°$/kbar, and for olivine, 34°/kbar.* The constant-volume slope greatly exceeds that of the melting curve, and melting will

---

*Values of compressibility and thermal expansion at 20°C used are:

Diopside:	β	$= 1.07 \times 10^{-6}$/bar (Adams and Williamson, 1923)
	α	$= 24 \times 10^{-6}$/deg (Kozu and Ueda, 1933)
Olivine:	β	$= 0.82 \times 10^{-6}$/bar (Bridgman, 1948)
	α	$= 24 \times 10^{-6}$/deg (Skinner, 1962)

take place if the system is confined to a fixed volume. It would be interesting to know the effect, if any, of stress on the melting curve: no data on any relevant systems are known to the writer. In a pioneering study, Coe and Paterson (1969) have demonstrated that the α–β quartz transition is raised by uniaxial compression with or without a confining pressure. That technique appears to be available for investigating a melting curve under similar conditions.

ADHESIVE STRENGTH OF CRYSTALS

In view of the importance of the tensile strength of rocks to many economic endeavors such as mining, it is surprising that a greater effort has not been made to understand why dissimilar minerals adhere to each other. The coherent and incoherent boundaries formed on exsolution of similar structures are now being studied (Yund, 1974), yet only one study has been made to evaluate the adhesive strength of dissimilar crystalline center forces. A quartz–feldspar interface was pulled apart in the absence of a confining pressure with a force of 60 to 100 bars, according to the tests of Savanick and Johnson (1972). They pointed out that, judging from the properties of the broken surfaces, the forces responsible for the adhesion operated over only a portion of the interfacial area. A more extensive study should be undertaken involving more parameters.

ESCAPE FROM THE CONTAINER

The present view of the initial liquid, then, is of drops at four-phase intersections, veins along three-phase boundaries, and films between grains. If the rock is stressed, the liquid aggregates along those boundaries perpendicular to the least compressive stress. The problem then seems to be how to assemble the aggregates into an anastomosing system of rivulets eventually leading into a feeder dike.

It is held by some that the container could be viewed as a three-dimensional sieve holding the liquid in the interstices, the liquid being released by breaking the wires in the sieve. Cracking of a brittle container seems to be inappropriate to a model of a magma chamber at very high temperature and presumably in a plastic state. Ductile fracture, wherein the solid undergoes marked deformation before fracture, is frequently observed in the failure of metals and yet not well understood (Biggs, 1970, p. 1227): that process may be a suitable description of the failure of a magma chamber. The apparent contradiction between earthquakes suggesting brittle failure and magmas indicating fluid or plastic behavior is often resolved by noting that a brittle response obtains at relatively low temperatures under short-term, low-magnitude (~100 bars) stresses, whereas plastic response occurs in the same material at relatively high temperatures under long-term, exceptionally low-magnitude stresses. Analogy is often made to the behavior of pitch because it flows like a liquid under its own weight, yet, when hit sharply with a hammer, breaks with a conchoidal fracture.

How, then, is the liquid to be moved from or through the hot plastic region surrounding a magma chamber into the cooler brittle region? Observation of a moving lava flow where cracks develop at the frontal crust and fresh lava oozes out at temperatures as low as 900°C (Spry, 1962, p. 214) is adequate proof that liquid can escape through the plastic region under a suitable force.

PLASTIC ENVELOPE

Plasticity is defined as continuous deformation without limit, relatively independent of pressure and without fracture, under a reasonably constant stress that exceeds a critical value (Figure 9-8). The mechanisms involving intracrystalline plasticity depend on the diffusion of material in various ways. Specific mechanisms dominate in various temperature regions, stress conditions, and strain rates. The boundaries of the regions of deformation mechanisms are diffuse because of such factors as grain size and water pressure. A deformation mechanism map (Ashby, 1972) has been prepared for olivine (Stocker and Ashby, 1973, p. 409) and is reproduced as Figure 9-9. At low tempera-

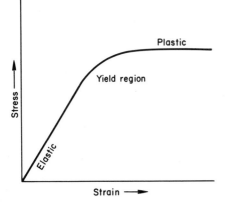

FIGURE 9-8 Description of the response of a rock to increasing stress depending on the observed strain for a constant strain rate.

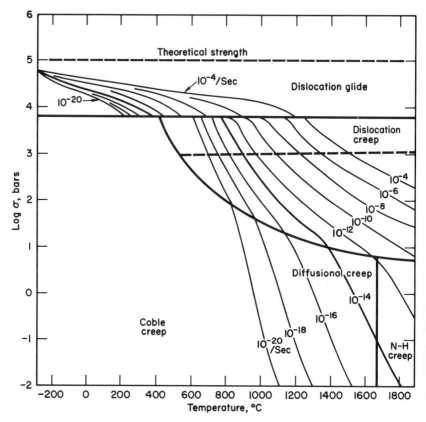

FIGURE 9-9   Steady-state deformation map for polycrystalline olivine (Fo_{85-95}) illustrating regions (outlined by heavy lines) of dominant deformation mechanisms as a function of the stress (σ) and temperature for various strain rates (light lines). High-stress dislocation creep regime occurs above the heavy dashed line. N–H = Nabarro–Herring. Grain diameter = 1 mm. Total hydrostatic pressure = 10 kbar. Activation volume for diffusion = 50 cm^3/mol (*cf.* atomic volume = 31 cm^3/mol). From Stocker and Ashby (1973, p. 409, Figure 7). (With permission. Copyrighted by the American Geophysical Union.)

tures and strain rates, the dominant deformation mechanism is "dislocation glide"; at intermediate temperatures and strain rates, it is "dislocation creep"; and at high temperatures and strain rates, it is lattice diffusion, "Nabarro–Herring creep." At temperatures somewhat below the melting temperature and at slow strain rates, the mechanism is grain boundary diffusion, "Coble creep." If the most likely conditions around a magma chamber are relatively high temperatures, slow strain rates, and low stress, the mechanism will be either Coble or Nabarro–Herring creep; if the stress is high, however, dislocation creep becomes important.

TIME DEPENDENCE

Strain rates for surface displacements are generally in the range 10^{-12} to 10^{-15}/sec, according to H. C. Heard (personal communication, 1975), based on geodetic measurements on faults (Whitten, 1956), rebound rates from crustal unloading (Crittenden, 1966), sea-floor displacement rates (Heirtzler *et al.*, 1968), and crustal shortening in orogenic regions (Gilluly, 1970). Unfortunately, most laboratory experiments are restricted to strain rates greater than 10^{-8}/sec, and extensive extrapolation is required for application of the results to the earth. If the effective viscosity and the stress of the plastic region are known, an estimate of the strain rate can be made through the relation

$$\eta = \sigma \dot{\epsilon},$$

where η is the viscosity; σ, the stress; and $\dot{\epsilon}$, the strain rate. For $\eta = 10^{10}$ poises and $\sigma = 0.5$ kbar ($= 5 \times 10^8$ dyne/cm^2), the strain rate is about 5×10^{-14}/sec. For an average upper mantle with $\eta = 10^{22}$ poises and $\sigma = 0.5$ kbar, the strain rate is 5×10^{-15}/sec, in accord with surface displacement rates due to isostatic rebound.

The response of the contained liquid to the strain rate in the host rock is equally important. The effective viscosity of basaltic magma at the surface is about 500 poises at the liquidus (Shaw *et al.*, 1968). Preliminary experiments by Kushiro *et al.* (in press) suggest that the viscosity is about 30–50 poises at 1375–1400°C and 20 kbar! For such an effective viscosity the liquid would literally squirt out of a containing rock under stresses of a few bars after an interconnecting liquid network is established. Furthermore, dissolved volatiles lower the viscosity markedly for tholeiitic melts but only negligibly for alkali olivine basalts (Scarfe, 1973, p. 101).

"KNEADING OUT" THE LIQUID

The plastic deformation of lherzolite has been studied by Carter and Ave'Lallemant (1970) at a series of temperatures (Figure 9-10). Their data show that at temperatures within the proximity of the solidus, lherzolite behaves plastically under very small stress differences. Preliminary data by the same authors show that even lower stress differ-

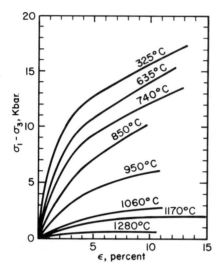

FIGURE 9-10   Deformation (ε) of a lherzolite under a confining pressure of 10–20 kbar and a strain rate of 7.8 × 10⁻⁵/sec as a function of differential stress and temperature. From Carter and Ave'Lallemant (1970, p. 2193, Figure 7A). (With permission of the Geological Society of America.)

ences are required if water is present. It appears that liquid could readily flow along intergrain boundaries until it reaches a region that can sustain brittle failure. In view of the ease of movement of liquid through the "kneading out"* process, it is reasonable to suppose that magma is more commonly extracted by fractional melting than by batch melting. In such cases it would be necessary to store the magma in auxiliary reservoirs to make large volumes available because the initial stages of aggregation depend on relatively slow diffusional processes. The behavior of lherzolite also suggests that mass movement of a mush, once formed, would quickly bring about liquid separation (Sleep, 1974). The source area appears to be best described initially as a region mainly crystalline with a continuous network of films of liquid, and yet it is difficult to abandon completely the notion of magma-filled caverns, albeit multiple and deformable, for auxiliary reservoirs.

If the spacing of volcanoes is controlled by the size and shape of the source region (see Chapter 10), the degree of melting of the source rock

---

*The "kneading" process was described by Chamberlin and Salisbury (1909, pp. 579 and 629–633) as the result of stresses arising from lunar and solar tides. Harker (1909, p. 323 *ff*.) described a similar process for the expulsion of *residual* magma under mechanical stress as the "straining off" or "squeezing out" of liquid. Later, the terms "wine-press differentiation," "filter pressing," and "filtration differentiation" were applied to the removal of residual liquids. The terms would be equally appropriate for initially formed liquids.

is small, and the liquid has a small viscosity, then the magma that feeds a specific vent flows from considerable distances. The nature of the flow is probably much more complex than flow of the Poiseuille type (Darcy's law) or of the Knudsen type, depending on the size of the capillaries. To the best of the writer's knowledge, the analysis of the flow of fluids through a stressed, porous, heterogeneous solid at high pressures and temperatures has not as yet been undertaken.

SOLUBILITY UNDER STRESS

Plastic flow involves recovery processes such as solution and re-precipitation as well as deformation mechanisms (twinning, gliding, and volume diffusion). The solubility of a crystal under a compressive stress is greater than it is under only hydrostatic pressure. This concept was first suggested (but not necessarily demonstrated) by Thomson (1862, p. 396), who applied a piston load to NaCl crystals immersed in a saturated solution. The compaction of the salt crystals was interpreted by Thomson to indicate that the salt was dissolved away at the stressed faces and deposited at the free faces. The circumstances of the Thomson experiment are analogous to those for crystals immersed in a magma saturated with those crystals.*

Unfortunately, consideration was not given to the effects of annealing recrystallization in the absence of a saturated salt solution, nor was it suggested how the stressed face, isolated from the salt solution by the piston face and bottom of the container, communicated with the ambient salt solution. A similar experiment was performed by H. K. Mao and P. M. Bell (personal communication, 1974) with gypsum in a saturated solution at very high pressure in a diamond cell, and no solution and redeposition were observed. The "Thomson effect" was qualitatively discussed by Sorby (1863, 1879) and applied to geological problems by Riecke (1895, 1912). A thorough review of the thermodynamics of recrystallization of stressed solids in the absence of a saturated solution was given by Paterson (1973). Although there appears to be a sound theoretical basis for enhanced solubility under stress (Mao and Bell, 1971), definitive experimental demonstration has not been achieved.

---

*The solubility curve is the same curve that describes the melting behavior of a crystal under hydrous conditions. The curve is continuous or discontinuous, respectively, depending on the absence or presence of critical end-point phenomena. In other words, the solubility curve describes the behavior for dilute solutions, whereas the melting curve describes the behavior for concentrated solutions.

LOSS IN STRENGTH OF CONTAINER

The presence of a melt within a rock also brings about a great loss in strength. In general, an increase in confining pressure produces an increase in strength for a wide range of rock types. The effect of confining pressure on the strength of basalt (diabase) and peridotite is given in Figure 9-11; these curves are typical in form for most rocks (Mogi, 1966).

With the formation of the first droplets of liquid in a rock, a pore

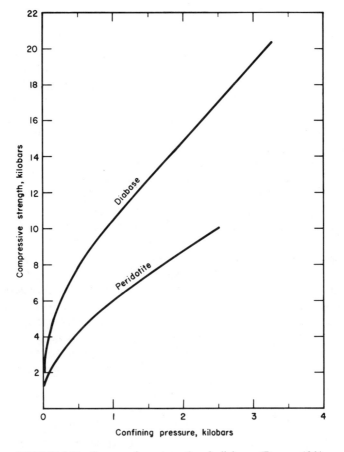

FIGURE 9-11   Compressive strength of diabase (Brace, 1964, p. 138) and peridotite (Mogi, 1965, p. 371) as a function of confining pressure.

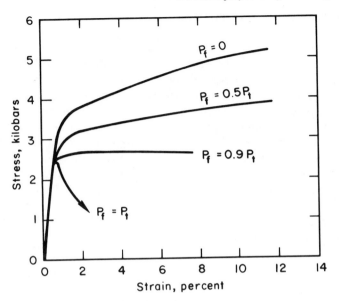

FIGURE 9-12 Stress-strain curves showing the effect of pore fluid (H_2O) on the wet strength and mode of deformation of a rock (Solenhofen limestone) at various ratios of pore-fluid pressure (P_f) to confining pressure (P_t). $T = 20°C$; effective porosity = 5.3 percent; grain size = 5.20 μm; $\sigma_3 = 3$ kbar. From Rutter (1972, p. 21, Figure 4A). (With permission of the Elsevier Scientific Publishing Company.)

pressure is generated. In effect, the pore pressure counters the confining pressure, thereby reducing the strength of the rock (Figure 9-12). The concept was demonstrated in detail by Hubbert and Willis (1957), Hubbert and Rubey (1959), and Rubey and Hubbert (1959) and experimentally investigated by Heard (1960), Handin *et al.* (1963), Raleigh and Paterson (1965), and Heard and Rubey (1966).

MAGMAFRACTING

Application of the pore-pressure principle to partial melting appears to have been made first by Mogi (1966), and to hydrothermal ore deposits by Phillips (1972). The propagation of cracks by increasing the pore pressure artificially, hydrofracting, has been used for many years, since 1948, for the purpose of increasing the permeability in oil reservoirs. It is appropriate to apply the term *magmafracting* to the effective decrease in strength of a rock by virtue of the generation of magma and the process of crack propagation by magma in the brittle failure region.

The propagation of the crack is again parallel to the maximum principal stress at depth; however, it is generally perpendicular to the least principal stress under relatively near-surface conditions. The common occurrence of vertical or inclined dikes at depth and the shallow formation of sills supports this view. The rate of propagation of cracks in polycrystalline materials at high temperatures is under current investigation, especially by ceramics engineers (Wachtman, 1974). Theories that include chemical reaction enhancement terms have been evolved for slow crack propagation (Wiederhorn and Bolz, 1970). Of particular interest is the experimental correlation of acoustic emissions with crack velocity in materials such as polycrystalline corundum at high temperatures (Evans *et al.*, 1974). Comparison of such emissions with earthquake foreshocks may be appropriate.

VOLCANISM PRODUCES SOME EARTHQUAKES

Dilatancy has been recorded in pre-earthquake episodes lasting 1–3 months and is potentially a tool for earthquake prediction (Nur, 1972).

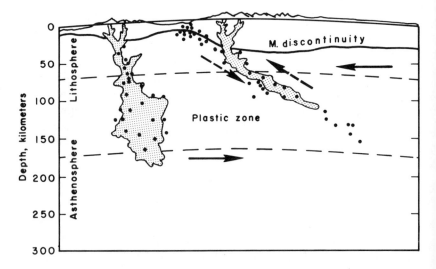

FIGURE 9-13　Plastic zone between lithosphere and asthenosphere as visualized by Anderson (1962, pp. 52–53). Black dots are earthquake foci. Stippled areas are magma. Arrows indicate relative motions of regions. (With permission of Scientific American, Inc.)

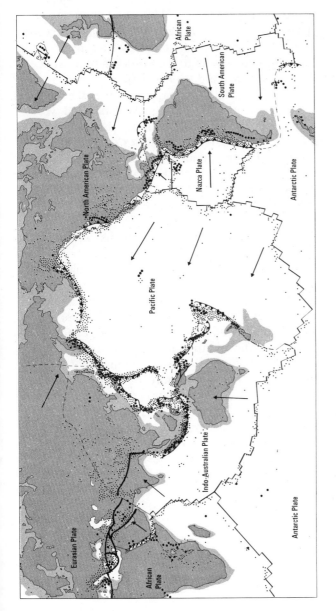

FIGURE 9-14 Location of volcanoes (heavy dots) relative to earthquake zones (light dots). Plate boundaries with offsets by transform faults are shown by solid and dashed lines; subduction zones, by line with triangles. Arrows mark motions of plates. Geological Museum of London (1972, p. 13, Figure 21). ("Crown copyright. Institute of Geological Sciences illustration, reproduced by permission of the Controller, Her Majesty's Brittanic Stationery Office." Fee paid.)

181

The increase in pore volume was believed by Nur to provide access for ground water, but could equally well provide access for magma. The presence of liquid decreases the strength of the rock, and rupture takes place. It would appear that magma may invade a region quiescently several months before its presence is announced by rupture. Near the surface, such dilatancy may contribute to the observed tumescence of the volcanic edifice (Mogi, 1958; Eaton, 1959, 1962).

Where conditions are appropriate for magmafracting, the aggregation of magma will, no doubt, produce earthquakes. It is the writer's view that earthquakes associated with volcanism are the result, not the cause, of volcanism. This view rests on the theses developed above: (a) the thermal regime outside the boundary of the magma chamber, marked by its solidus, suggests plastic rather than brittle behavior of the confining rocks; and (b) the aggregation of the magma into rivulets and cracks occurs through the process of magmafracting in the region of brittle behavior.

In Anderson's view (1962), the plastic zone is not just a local region around a magma chamber but is a broad (60–250 km thick), ill-defined region of low velocity marking the transition from the lithosphere to the asthenosphere (Figure 9-13). The temperature in the plastic zone approaches the melting point of the rock, and magma appears where zones of disturbance are marked by earthquakes. There is indeed strong correlation of earthquakes with volcanic eruption (Figure 9-14). It is evident that the deformational structures of a region control the position of the volcano at the surface. The regional fabric of the prestressed environment exerts the major control on the location of new volcanoes; Fiske and Jackson (1972) demonstrated that the position of subsequent volcanoes may be the result of the gravitational stresses in pre-existing edifices. There is considerable debate on whether or not earthquakes associated with volcanism record the strain released as magma rises to the surface. At present, the issue is not resolved; however, the commentary in the next chapter on the energetics of volcanism may be of some value.

10 Energetics and Periodicity

ENERGETICS

PARTITION OF ENERGY

The energy released in a volcanic eruption is both thermal and kinetic. Estimates of the amounts released have been made by summing the contributions by volcanic earthquakes, potential energy due to lifting of the lava, thermal energy released from the flows, losses due to heat conduction and gas expansion, thermal-spring activity, and kinetic energy of ash and pumice eruptions. Estimates of the total energy for the largest eruptions, calculated by Yokoyama (1957), are listed in Table 10-1. Only about half the magma generated reaches the surface in Hawaii, according to D. A. Swanson (personal communication, 1975), and the wave energy received at teleseismic distances is probably only a few percent of the total yield (Bath, 1966, p. 161). The inferred relations for Mount Rainier, Washington, prepared by Fiske *et al.* (1963), illustrate why a large proportion of the magma produced does not reach the surface (Figure 10-1).

The partition of released energy in explosive events may be crudely related to possible kinetic–thermal conversions in the alleged generation of magmas by seismic events. According to an analysis by Gault and Heitowit (1963), the energy in a hypervelocity impact in basalt is about 19–23 percent thermal, the remainder being distributed between comminution and ejecta, with a negligible amount lost to compaction.

183

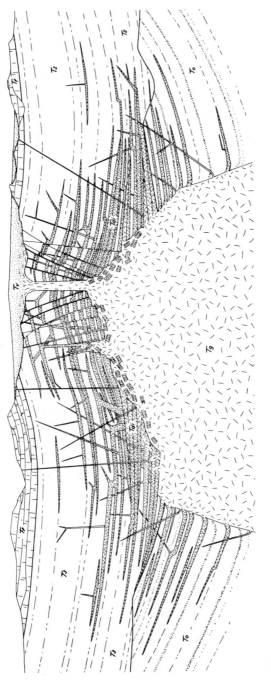

FIGURE 10-1 Cross section of Mount Rainier, Washington, illustrating the large volume of magma (T_g) that penetrates the volcanic edifice as sills (T_{di}) and dikes and does not erupt. The contemporaneous erupted rocks are labeled T_v. T_o, T_s, and T_f are various Tertiary formations of volcanic products, including mudflows and siltstones. From Fiske *et al.* (1963, p. 51, Figure 38). (With permission of the Director of the U.S. Geological Survey.)

TABLE 10-1 Estimate of Total Energy Released in Dated Volcanic Eruptions[a]

Volcano	Year	Energy Released (erg)
Tambora	1815	8.4×10^{26}
Sakurajima	1914	4.6×10^{25}
Bezymianny	1955–1966	2.2×10^{25}
Krakatoa	1883	ca.1×10^{25}
Asama	1783	8.8×10^{24}
Fuji	1707	7.1×10^{24}
Capelinhos	1957	4×10^{24}
Sakurajima	1946	2.1×10^{24}
Kilauea	1952	1.8×10^{24}
Torishima	1939	9.7×10^{23}
Komagatake	1929	5.6×10^{23}
Miyakeshima	1940	4.8×10^{23}
Bandaisan	1888	ca.1×10^{23}
Pematang Bata	1933	4.5×10^{22}
Una-Una	1898	1.8×10^{22}
Mihara	1954	1.3×10^{22}
Arenal	1968	1×10^{22}
Adatarasan	1900	6.4×10^{21}
Asama	1938	4.0×10^{21}
Mihara	1912	6.3×10^{20}
Tokachidake	1926	2.8×10^{20}
Showa Sin-zan	1944	1.4×10^{20}
Kusatsu-Shirane	1932	1.6×10^{18}

[a]After Yokoyama (1957, p. 96, Table V) and Macdonald (1972, p. 60, Table 4-3).

The thermal energy in a contained underground nuclear blast would be expected to be somewhat greater. Butkovich (1974) estimated the amount of melt generated by an underground nuclear explosion to be about 1,000 metric tons of melt per kiloton of energy yield in a rock with a density of 3.0. The energy released by 1–100 megatons of TNT explosive is about equal to that of large volcanic eruptions (10^{22}–10^{25} erg). If the energy yield was 10^5 megatons (10^{27} erg), the amount of melt produced would only be about 36 km^3. Such an energy yield is equal to that in an earthquake measuring 8.5 on the Richter scale, so it is not likely that shock waves of moderate earthquakes will produce large volumes of magma, even if the energy conversion is as great as that in a nuclear explosion.

ENERGY IN EARTHQUAKES

The energy released in earthquakes can be estimated from the empirical relation of Gutenberg and Richter (1949, p. 19):

$$\log E = 12 + 1.8M,$$

where E = energy released in ergs and M = magnitude of earthquake as defined by Richter (1935).[*] A magnitude-6 earthquake releases about 10^{23} erg, whereas the greatest earthquakes on record, $M = 8.6$ (e.g., near Honshu, Japan), released 10^{28} erg. The rough equality of the energy released in a volcanic eruption and in a moderate earthquake is presumed to be fortuitous. The efficiency of conversion of mechanical energy into thermal energy is no doubt low, and earthquakes are related primarily to the movement of magma rather than its generation. The relatively close association in time of some larger earthquakes with eruption supports this view (Figure 10-2).

Several attempts have been made to relate surface fault length to the magnitude of an earthquake (Tocher, 1958; Iida, 1959; Press, 1965). Such an empirical relation might be useful in estimating the propagation of a magma towards the surface. Unfortunately, the wide variations in stress drop, source dimensions, and the complexities of measuring the parameters of the fault itself have not yielded a reliable relationship (Brune and Allen, 1967, p. 511).

FORCES CAUSING MAGMA RISE

OVERBURDEN SQUEEZE

What, then, are the forces that cause magma to rise to the surface? Two of the major processes are (a) overburden squeeze and (b) buoyancy. The main force probably results from the difference in pressure generated by the rock column and that generated by the

---

[*]The constants of the equation have changed with time and not all the data fit a linear form. Some seismologists prefer the equation

$$\log E = 11.8 + 1.5M,$$

based on Gutenberg and Richter (1956). It should be evident that the calculation of energy from seismic waves will undergo further revision as the parameters of a shock are more fully understood.

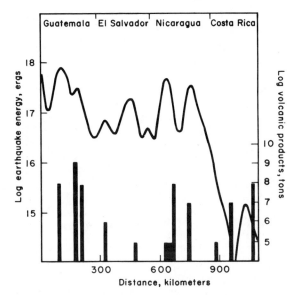

FIGURE 10-2 Smoothed energy released (left scale) for the period 1963–1971 from intermediate-depth (70–150 km) earthquakes versus distance along the Central American arc. Vertical bars are orders-of-magnitude estimates of tons of volcanic products erupted (right scale) during same period at active volcanoes. From Carr and Stoiber (1973, p. 333, Figure 2). (With permission of *Bulletin Volcanologique*.)

magma column. The hydrostatic pressure at the base of a column may be calculated from

$$P = \bar{\rho}gx,$$

where $\bar{\rho}$ = average density, g = acceleration due to gravity, and x = depth. For ρ_{rock} = 3.00 and ρ_{magma} = 2.78, the hydrostatic pressure varies as shown in Figure 10-3A. At 60 km the pressure at the base of the rock column is 18.0 kbar and that at the base of the magma column is 16.7 kbar. The effect was displayed by Holmes (1945, p. 478), and a similar diagram is given in Figure 10-3B. To equalize the pressure at the base of the columns, the magma must rise 4.7 km ($\approx$15,340 feet)* above the surface. The highest volcanoes (not necessarily basaltic) rarely

*The highest continental extinct volcano in the world at present is Cerro Aconcagua, Argentina, with an elevation of 6.96 km (22,834 feet). It is more than 11.83 km (42,824 feet) above the Peru–Chile trench. The highest volcano measured from its submarine base (in the Hawaiian trough) to its peak is Mauna Kea, Hawaii, with a total elevation of 10.20 km (33,476 feet).

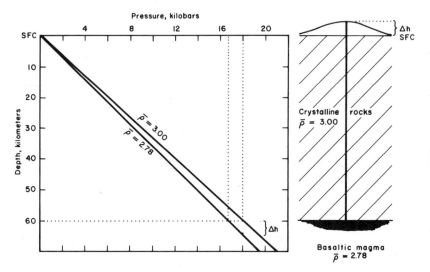

FIGURE 10-3   *A*, Relationship of hydrostatic pressure to depth in a column of magma (average density, $\bar\rho = 2.78$) and a column of rock ($\bar\rho = 3.00$). *B*, Schematic cross section of a volcano, illustrating the rise of magma (Δh) above the reference surface (SFC) necessary to equalize the pressure of the rock load on the magma chamber and the magma column.

exceed this height. The implication that magmas are generated or separated at a depth of about 60 km cannot be ignored; however, there are many factors, most of which have not been evaluated quantitatively, that would have to be considered before such a limit on magma generation were applied.

BUOYANCY

An additional force in causing magma to rise to the surface is buoyancy. The velocity of rise of a sphere of melt is given by Stokes' law:

$$ v = \frac{2ga^2\,(\rho_1 - \rho_2)}{9\eta_{\text{med}}}, $$

where g = acceleration due to gravity, a = radius of sphere of melt, ρ_1 = density of medium, ρ_2 = density of melt, and η_{med} = viscosity of the medium. For a sphere of melt with $a = 1$ km (10^5 cm), $\rho_1 = 3.4$ g/cm^3, $\rho_2 = 2.8$ g/cm^3, $g = 10^3$ cm/sec^2, and $\eta_{\text{med}} = 10^{19}$ poises, then $v = 0.13 \times 10^{-6}$ cm/sec. The velocity is about 4 cm/yr, the average rate of plate spreading at the midocean ridges. The implication is that buoyancy could account for the rise of sufficient magma to satisfy the present annual worldwide production.

The buoyant rise of magma has been related to the growth of salt domes (e.g., Bayly, 1968, p. 117). Salt domes have long been considered products of a diapiric process (Nettleton, 1936); a new analysis by R. O. Kehle (personal communication, 1975), however, revives the suggestion that overburden squeeze (Barton, 1936) may be responsible for their puncturing overlying sediments. The pressure generated by the lithostatic loading of a delta, for example, is transmitted through underlying plastic salt to that part of the salt bed covered only by thin sedimentary rocks, presumably in deeper water. The density contrast between the deltaic material and ocean water is apparently sufficient to generate a pressure difference adequate for extruding the more thinly covered salt as domes.

Stokes' law can also be used to calculate the velocity required to bring up nodules observed in alkali basalts. If the effective viscosity of the magma is 10^3 poises and the maximum radius of a nodule is 2.5 cm, then $v = 0.83$ cm/sec or about 3 m/hr. The rate of rise required to *support* the nodules in the magma is not great compared with the velocities observed for lava flows.

IMPORTANCE OF VISCOSITY

The viscosity of magma is obviously a most important parameter in many aspects of its aggregation and flow. The viscosity coefficient is the ratio of the shear stress to the strain rate, and when the ratio is a constant, the magma has the behavior of a Newtonian fluid (Figure 10-4). When the ratio is not a constant, the magma exhibits non-Newtonian behavior. If a critical value of the shear stress must be

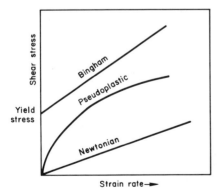

FIGURE 10-4 Schematic relationship between shear stress and strain rate for Bingham, pseudoplastic, and Newtonian magmas.

achieved before flow can take place, the magma is classed as a Bingham fluid. The behavior of a Bingham fluid is like that of house paint: It can be spread easily with a brush, but stays essentially in place without flowing when the brushing action stops. Shaw *et al.* (1968) found that the response of the lava in the Makaopuhi lava lake, Hawaii, was closest to that of a Bingham fluid with critical shear stresses of 700 and 1,200 dyne/cm^2 (0.7 × 10^{-3} and 1.2 × 10^{-3} bar, respectively) at two different depths in the lake. The respective viscosities at the two positions were 7.5 × 10^3 and 6.5 × 10^3 poises. The energy required to move magma of the Bingham type will be obviously greater than that to move the Newtonian type. The yield stress also varies with composition, according to Hulme (1974, p. 378). He indicated that basaltic magmas yield at a much lower value of stress than andesitic magmas (Figure 10-5). The loss of volatiles may have led to an overestimate of

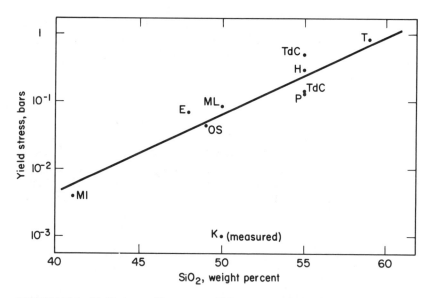

FIGURE 10-5   Yield stress of lava versus SiO$_2$ content calculated from depth of flow, slope, and density. Crystal and gas content and temperature are unknown.

E	= Etna, 1966	*ML*	= Mauna Loa, 1942
H	= Hekla, 1947	*OS*	= O-Shima, 1951
K	= Kilauea, measured (1130°C; ~25% crystals; 2–5 percent gas bubbles)	*P*	= Parícutin, 1945–1946
		T	= Teide
MI	= Mare Imbrium	TdC	= Tristan da Cunha

After Hulme (1974, p. 378, Figure 16). (With permission of The Royal Astronomical Society.)

the viscosities measured by Hulme on the basis of levee height, ground slope, and lava density.

Emphasis has been placed on the eutecticlike character of basaltic liquids, and it is not difficult to accept the idea of Feild and Royster (1918) that viscosity is minimum at a eutectic. With disregard for the structural change of liquid with composition, one might argue that the eutectic composition should have the maximum viscosity because it has the lowest temperature. Other studies (Herty *et al.*, 1930), however, suggest that the nature of the liquidus surface has no effect on the viscosity. A high yield stress in a liquid of the eutectic composition would indeed make the removal of small amounts of melt difficult.

PERIODICITY

The periodicity of eruption is one of the keys to predicting the outbreak of a volcano. Some workers have assumed that volcanic activity is a random process in both space and time (Verhoogen, 1946, p. 770; Wickman, 1966, p. 291). Other investigators believe there are specific patterns to volcanic activity in space and time. The great outpourings of basalt were associated by Du Toit (1937, pp. 94, 174) with the periodic breaking up of the lithospheric plates. On a shorter time scale, one investigator (Dubourdieu, 1973) examined the records of active volcanoes throughout the world using data from the sixth century forward and suggested that volcanic eruptions have the same frequency as seismic activity, approximately 4 years. McBirney *et al.* (1974) presented evidence for a possible 5-m.y. cycle in the Oregon Cascades and suggested that the "same episodes occurred in unison over a large part of the earth."

PAUCITY OF DATED BASALTS

The major difficulty in assessing periodicity is the small number of volcanoes studied. A small sample would tend to result in the description of volcanism as intermittent, episodic, or spasmodic. Some appreciation of the worldwide frequency of magma production may be gained by examining the ages of igneous rocks. Older basalts are dated with great difficulty, and only few data are available; however, the ages recorded in minerals from all rocks, although subject to some correction, may serve the purpose (Figure 10-6A). A crude periodicity of magma generation is evident, and it is assumed that basaltic magmas are also generated in large amounts over specific time periods. An estimate by Engel *et al.* (1965) of the periodicity and volume of basalts

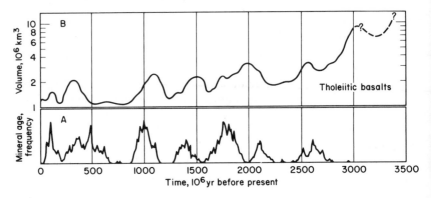

FIGURE 10-6   A. Frequency of mineral "age" from 0 to 3000 m.y. before present from worldwide localities, compiled by Gastil (1960, p. 5, Figure 1). (With permission of the *American Journal of Science*.)
    B. Volume of continental tholeiitic basalt extruded worldwide, estimated from geological mapping by Engel *et al*. (1965, p. 729, Figure 4). (With permission of the Geological Society of America.)

based on geologic mapping is given in Figure 10-6*B*. Menard (1964, p. 87 *ff*.) called attention to the 10,000 or more volcanoes in the Pacific Basin, most of which lie in the western or southwestern sector, emphasizing their maximum growth in periods during the late Mesozoic through early Cenozoic and again from the late Tertiary to the present. The duration of a single volcano is probably in the range 1–5 m.y.; however, a magma source may be of much longer duration, according to the heat- and mass-transfer calculations of Shimazu (1961).

SPACING OF VOLCANOES

Some volcanoes, especially in island arcs, also appear to exhibit a periodicity in spacing. Marsh and Carmichael (1974, p. 1202) plotted the distance between neighboring volcanic centers along the Aleutian Islands, Alaska Peninsula, and Cascades and obtained a spacing of about 70 km (Figure 10-7). They attributed the spacing, after Ramberg (1972), directly to the growth of perturbations on a gravitationally unstable layer deformed in a sinusoidal fashion (Figure 10-8). The size of the conduits is probably of the order of 25 km in diameter (Koyanagi and Endo, 1971).

    Based on laboratory experiments, Ramberg (1972, p. 52) believed that the critical feature causing the episodic character is that the rate of

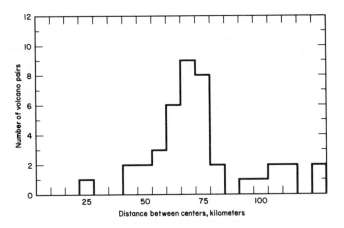

FIGURE 10-7 Histogram of distance between pairs of volcanic centers in the Aleutian Islands, the Alaska Peninsula, and the Cascades. From Marsh and Carmichael (1974, p. 1202, Figure 8). (With permission. Copyrighted by the American Geophysical Union.)

ascent of the perturbation in the buoyant layer increases at a rate in excess of a linear rate with the thickness of the layer. In other words, as the buoyant layer thickens with continued heating, for example, the rate of rise increases and the sinusoidal form amplifies rapidly. After the diapir is detached by some unspecified process, there is a period of quiescence until sufficient heating produces a renewed thickening of the layer undergoing melting, and the process repeats.

Spacing of volcanoes may be controlled at depth by the size and shape of the source region of each vent. For an extreme example, the volume of the Mauna Loa, Hawaii, edifice is about 10^5 km^3, and the source region will be 3.3×10^5 km^3 if the magma type was generated from 30 percent partial melting. If the magma was produced from 5 percent partial melting, then 2×10^6 km^3 of parental material will be

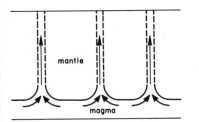

FIGURE 10-8 Schematic cross section of mantle, illustrating gravitational stabilization of a low-density melt by forming equally spaced diapiric conduits. From Marsh and Carmichael (1974, p. 1202, Figure 9). (With permission. Copyrighted by the American Geophysical Union.)

involved. The shape of the source region is obviously unknown; however, if the magma is withdrawn from a cylindrical region 50 km thick, then the nearest volcano of equivalent dimensions will be about 90 km away for 30 percent partial melting and about 225 km away for 5 percent partial melting. Although no overlap is assumed, continued melting of an already partially depleted region is more likely. The average spacing determined by Marsh and Carmichael, 70 km, implies that cylindrical source areas would have to be 8.7 or 52 km thick for partial melting of 30 percent or 5 percent, respectively, for volcanoes with an average volume of 10^4 km³. Marsh and Carmichael (1974, p. 1204, Figure 10) envisaged a *horizontal,* cylindrical source region of small diameter supplying the vents. Cylindrical source areas are not considered to be very realistic, and the reader may wish to try other dimensions and geometric shapes of source regions more compatible with his preferred mechanism of magma generation.

AGE AND SPACING RELATED?

The evolution of the Hawaiian Ridge also appears to be episodic in space and time. Jackson *et al.* (1972) have shown quantitatively that the Hawaiian volcanoes increase in age from east to west and are spaced in a systematic way (Figure 10-9). An early idea attributed the southeastward extension of the Hawaiian chain to a slowly propagating fracture that intercepted new regions of undepleted mantle undergoing

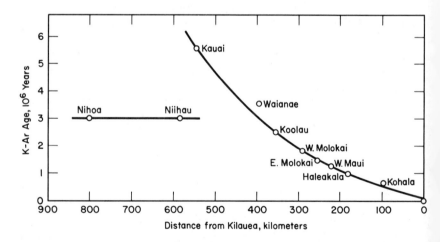

FIGURE 10-9   Age (K–Ar) of tholeiitic volcanism versus distance from Kilauea for the Hawaiian chain. Lines connect ages of adjacent shields. From Jackson *et al.* (1972, p. 608, Figure 3, inset). (With permission of the Geological Society of America.)

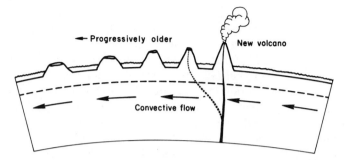

FIGURE 10-10   Schematic cross section of volcanic island chains illustrating generation of a new volcano as a result of the motion of the oceanic plate relative to the magma source. After Wilson (1963, p. 93). (With permission of Scientific American, Inc.)

partial melting (Betz and Hess, 1942, p. 109 *ff.*). With the advent of plate tectonics, Wilson (1963) suggested that sequential eruptive activity was due to the passage of the Pacific Plate over a "hot spot" in the mantle from which the magma was derived (Figure 10-10). Jackson and Wright (1970, p. 426) argued that the hot spot would soon be depleted, and the relative constancy of magma composition could not be explained. Therefore, they suggested that the source as well as the plate moved and accepted the idea of a slowly propagating fracture. The hot spot was viewed by Morgan (1971) as a "plume" of hot material rising from the mantle. Jackson *et al.* (1972) defined a "melting spot," without regard to process, of about 300 km in diameter from which tholeiitic magma could be supplied to the moving plate as the spot itself progressed.*

Many of the seamount volcanoes in the Pacific Ocean appear to occur in groups of about 10 to 100, and the center of volcanism usually migrates along a lineation (Menard, 1964, pp. 76–79). The sequential character of eruption is considered a general feature, and relative motion, whether mainly in a propagating mantle source or mainly in the lithosphere plate, seems to play a major role.

CONVECTIVE HEAT TRANSFER MODELS

JOLY–COTTER MODEL

One of the earliest ideas on the periodic nature of magmatism was proposed in 1923 by Joly (1930, pp. 192–193) and elucidated by Cotter

*The terms "plume" and "melting spot" are closely analogous in concept to the terms "ascending column" and "hot spot" of Holmes (1945, p. 410, 483).

(1924). Joly considered the parental material to be crystalline basalt, its melting curve being intersected by the geothermal gradient at a certain depth. As the basalt melts, a quiescent period ensues because of the time required to obtain the enthalpy of melting. After reaching an "advanced stage" of melting, the weakened material breaks down from "the disruptive action of tidal forces" and convection begins. As the melt rises adiabatically, its temperature is above the melting point appropriate for the reduced pressure, and the liquid begins "to melt away and to erode the solid rock above it. . . . " The molten zone is advanced upward as the descending cooled melt crystallizes at the bottom of the liquid mass. The rise of the molten zone is terminated mainly by heat losses to the surface due to thermal conduction from the molten zone. After a sufficient period of radioactive heating, thermal conditions are restored, the basalt is again brought to the melting temperature at depth, and the cycle is repeated. Joly's theory does not depend on the starting material's being basaltic, but it would fail if a more refractory layer intervened. If the starting material were garnet peridotite, any zone melting process would have to be inefficient so that basaltic fractions would remain after each successive pass.

TIKHONOV ET AL. MODEL

Joly's theory for thermal cycles was reconsidered by Tikhonov *et al.* (1970), who emphasized the high heat transfer in the molten layer itself. In one numerical model, they suggested that radioactive heating of the earth produced the first primordial melt at a depth of 460 km about 1.8–2.4 b.y. ago (Figure 10-11). The number of periods is four, and the mean duration is about 300–500 m.y. With different heat-transfer rates and thermal conductivities, a greater number of periods were obtained in other models. With suitable choice of parameters, a model could be constructed to fit the seven (?) periods of worldwide volcanism beginning about 3.5 b.y. ago with a mean interval of about 400 m.y.

ROLE OF VOLATILES

Volatiles have always been high on the list of factors believed to be responsible for the periodic character of volcanism. The development of pressure in magmas as a result of crystallization was considered by Morey (1922), on the basis of his studies in the $K_2O–SiO_2–H_2O$ system, to be the cause of the paroxysmal eruptions, especially of salic magmas. The key point of his argument was that in the presence of an excess of volatiles the pressure rises with cooling along the univariant

Time, 10^9 yr before present

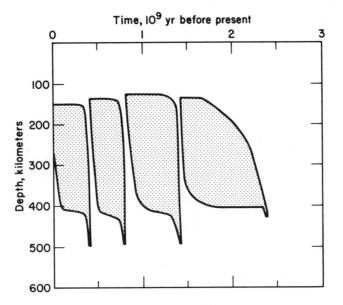

FIGURE 10-11 Melting periods in the mantle as a function of depth, according to the mathematical treatment of Tikhonov *et al.* (1970, p. 329, Figure 3). The calculation involves a solution for two moving boundaries where one interface is melting and the other is crystallizing. Intensive heat transfer is assumed in the melted layer. The lower interface "catches up" with the upper interface at 130–150-km depth. The four periods vary in mean duration from 300 to 500 m.y. (With permission of Elsevier Scientific Publishing Company.)

curve describing the equilibria between crystals, liquid, and gas (Figure 10-12). It is necessary, therefore, for a magma to reach a condition of univariance before high pressure can develop and "break" the chamber. The resealing of the chamber after eruption and continued cooling again causes the gas pressure to build up, and the chamber is "fractured" with the extrusion of the magma. The repetition of these events was believed to be one of the main causes of periodic eruption with relatively short repose times. The large number of components in a magma makes the attainment of such a univariant condition unlikely (Yoder, 1958). An additional difficulty is that high-pressure phases such as pyrope (Figure 10-12) and jadeite (Boettcher and Wyllie, 1969, p. 899, Figure 9) have a positive dT/dP in the presence of excess H_2O. That is, pyrope and jadeite cool along the hydrous melting curve with *decreasing* pressure. [It should be recalled that melting of the common

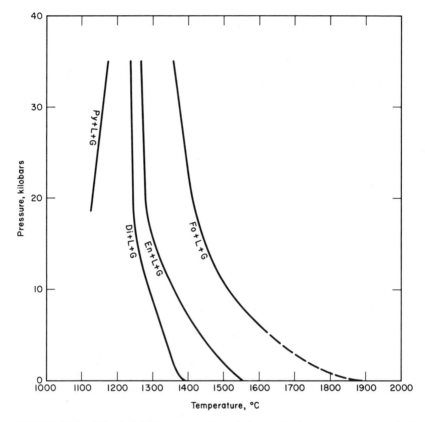

FIGURE 10-12   Effect of H_2O on the melting of the principal end-member minerals in garnet peridotite.

Forsterite (Fo)–H_2O: Kushiro and Yoder (1969, p. 155, Figure 52).
Enstatite (En)–H_2O: Kushiro et al. (1968b, p. 1690, Figure 2); coexisting forsterite ignored.
Diopside (Di)–H_2O: Yoder (1965, p. 87, Figure 12); Eggler (1973, p. 458, Figure 17).
Pyrope (Py)–H_2O: Mysen (1975; unpublished data).

silicates in the presence of a fixed partial pressure of water generates a liquidus with a positive dT/dP, whereas the solidus remains the same as in the presence of excess water (see Figures 4-16 and 4-17).] This relationship is also observed at high water pressures in natural spinel lherzolites (Figure 4-20).

Explosive volcanism is more likely the result of a magma reaching saturation with regard to volatile components. Saturation may be reached as a result of crystallization, the gas content of the magma increasing as crystals form. (If hydrous phases are formed, the water content of the magma may increase or decrease, depending on the initial amount of water.) In addition, Yoder (1965) illustrated with the experimental results on the diopside–anorthite–H_2O system that release of pressure may also cause an undersaturated magma to reach saturation. The liberated gas phase, requiring a much larger volume than when dissolved, acts as the propellant in the explosive ejection of a magma exsolving gas. The vesiculation of the magma may take place quite independently of the crystal content, yet crystals may in fact accelerate the vesiculation process. It is not necessary, therefore, for a magma to reach a condition of univariance for explosive eruption. Most basaltic magmas arrive at the surface quiescently, and the volatile content is exceptionally low. An explosive basalt (?) eruption such as that at Tambora, East Indies, in 1815 is indeed a rarity, and there volatiles acquired by the magma from the ground and sea were considered by some investigators to be the cause. It seems most unlikely that ground water or sea water would be absorbed by a hot magma— the water was probably a primary constituent of the magma. On the whole, the role of volatiles in the periodic eruption of basaltic lavas seems to be minor.

ROLE OF SUBSIDIARY RESERVOIR

The periodicity of eruptions within a single volcano suggests a process dependent on the refilling of a reservoir. Hekla in Iceland exhibits sudden bursts of activity, with changes of magma composition reflecting depletion of a reservoir (Figure 10-13). Mt. Vesuvius in Italy has similar behavior, and it also appears to have a constant rate of magma production (Wickman, 1966, p. 361). In other volcanoes having a subsidiary reservoir (e.g., Kilauea, Hawaii) the periodicity may be dependent more on the events taking place at the site of magma generation than on the physical properties of the subsidiary reservoir. It is not known whether all volcanoes have a subsidiary reservoir that governs the periodicity or whether the periodicity depends on the rate of production of magma at the source. The critical parameters have not yet been identified, but present opinion relates the short-term periodicity to the physical properties of the volcanic edifice that undergoes tumescence and rupture. Shaw and Swanson (1970, pp. 287–288) note, however, that the "viscous collapse" of a magma chamber does not appear to be rapid enough to maintain constant fluid pressure.

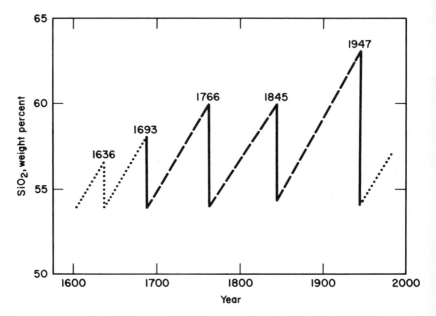

FIGURE 10-13 Relation of SiO$_2$ content of initial and final analyzed products of Hekla, Iceland, eruptions (solid lines). Quiescent intervals are represented by dashed lines. Dotted lines indicate periodicity based on prehistoric tephra layers and extrapolation. From Thorarinsson (1954, p. 43, Figure 7). (With permission of Societas Scientarium Islandica.)

RUNAWAY MECHANISMS POTENTIALLY PERIODIC

Mention has been made of the runaway character of shear melting and melting associated with a minimum in thermal conductivity. In each case the heating increases at an accelerating rate as the local viscosity or thermal conductivity decrease. The heating "runs away" until the energy source is depleted, convective heat transfer exceeds the production, or the liquid is removed from the generation region. After some period of quiescence the runaway character may again be initiated and the process repeated.

In summary, several mechanisms have been proposed that may become periodic locally; however, only the Joly–Cotter scheme, suitably adjusted for garnet peridotite, appears to provide for worldwide magmatic events with long-range periodicity. The catastrophic ruptures that lead to plate separations may result from the mantle convection evolving from worldwide magma generation. For this reason the principle of the Joly–Cotter model may be best suited to present-day concepts of periodic mantle dynamics.

11 Summary Overview

The opportunities for research on the problems of basaltic magma generation are indeed manifold. In view of the great need for fundamental data and methods to collect the necessary data, it is useful to present an overview of the present state of knowledge. Whereas an attempt to provide an evaluation at this stage of uncertainty may yield a prejudicial view, a framework is helpful in planning research. Adequate tests of conclusions soon indicate more appropriate directions to pursue.

WHAT IS MELTED TO YIELD BASALTIC MAGMA?

The seismic data indicate that the crust and mantle are crystalline for the greater part. The composition and mineralogy of that crystalline material depend considerably on how the earth formed. If the earth formed by the relatively cool accretion of condensed matter, differentiated relative to other planets and heterogeneous, then one may visualize a mantle consisting of masses, of a wide range of size, age, and composition, welded together by metamorphic processes. Melting due to the conversion, albeit inefficient, of gravitational energy to thermal energy may have occurred only in the core and lower mantle. On the other hand, the crust and mantle may have resulted from the differentiation of a molten mass, initially homogenized by convection.

201

The randomly heterogeneous earth and the continuously layered homogeneous earth represent two extreme distributions of mantle material, and many variations within these extremes have been assumed. Many of the subsequent concepts of magma generation implicitly and explicitly depend on the assumptions regarding the origin of the earth.

It was concluded that the assemblage of minerals in the mantle most likely to yield magma of basaltic composition is garnet peridotite; this mineral assemblage could vary widely in the proportions of phases but must necessarily remain restricted in the kinds of major phases present. The distinct petrographic provinces, that is, the variability in composition between regions, are attributed in part to the presence or absence of minor phases such as phlogopite as well as variation in the type of solid solution within the major phases of the mantle. Another major factor in the development of a petrographic province is no doubt the tectonic environment and the associated mechanism(s) for magma generation. It is not known whether parts of the upper mantle also differ in age as well as composition. The absence on the earth of a historical record from 4.6 to 3.8 b.y. leads one to depend on the interpretation of processes recorded on the moon. The severe impacting believed to climax at about 4.0 b.y. may have churned up the crust and mantle to the depths from which magmas are now believed to be derived. That is, early vertical differentiation or inhomogeneity in the upper mantle and crust may have been wiped out by the homogenizing effects of impacting.

It is not possible to specify which garnet peridotite is most representative of the upper mantle because the samples available are believed to be more or less depleted of basaltic constituents. Adequate tests of the degree of depletion have not been carried out; however, relatively lower Cr_2O_3/Al_2O_3 and higher FeO/MgO may be suitable guides for relatively undepleted parental material (Dickey and Yoder, 1972; Boyd and Nixon, 1975, p. 452). Furthermore, the relationship between derivative eclogitic liquids—that is, liquids that would crystallize to eclogite mineralogy—and garnet peridotite is obscure because the eclogite samples are not clearly distinguishable as cumulates, liquids, or remelted basalts. Attempts to relate basalts, as representative derivative liquids, to garnet peridotite fail because most basaltic magmas are believed to have undergone considerable change en route to the surface. Some ambiguity is introduced, therefore, when their relations at high pressure are analyzed. In addition, experimental data are lacking on the relevant partition coefficients of key elements at various pressures.

WHERE DOES MELTING OCCUR?

The drop in velocity and increase in attenuation of seismic waves suggest that some broad regions of the earth's mantle are in a partially (5–8 percent) molten state at depths of 70 to 150 km. Other regions exhibit possible molten regions at 300–400 km, whereas some old continental cratons and the older oceanic regions show no evidence of partial melting at depth. The shallowest regions of magma are *presumed* to occur under the midocean ridges, yet there is no seismic evidence. The initial site of generation may have been several hundred kilometers below, the liquid content of the magma increasing en route upwards.

Estimates of the local geothermal gradient are related to the beginning-of-melting curves of mineral phases common to garnet peridotite, and their intercept signifies a possible zone of melting. Unfortunately, it is necessary to specify the volatile content and composition in order to fix an appropriate beginning-of-melting curve. Vast regions of the mantle would be molten if H_2O were present in amounts in excess of that needed to saturate the liquid. Because the seismic evidence indicates only small amounts of partial melt where melting ensues, the volatile content is assumed to be low. The ratio of CO_2 to H_2O, for example, could be varied to achieve the desired amount of melting, but the depth range of melting is still considerable. The source of volatiles in the mantle and the extent of depletion in the early stages of ocean formation become important issues.

The concept of a geotherm is not a persuasive constraint after the processes of melting are examined. The ideas of convection and buoyancy, for examples, focus on the adiabat. The rise of a mass to lower pressures without loss of heat was shown to be adequate for melting. The depths where melting could begin appear to be confined not by fixed relationships between a geotherm and the melting curve, but by the intersection of the family of adiabats from the local thermal gradient with the melting curve.

HOW DOES MELTING BEGIN?

A wide array of mechanisms have been proposed for the initiation of melting. Concepts involving the relief of tensional or compressive stresses require stresses greater than those believed to exist in the mantle. For mechanisms depending on the conversion of mechanical energy to heat, a broad range of heat-producing capacity is assumed. Whatever the capacity, these mechanisms appeared to fail with the onset of melting. Brittle failure and high friction are not the properties of a rock mass near its melting temperature.

Places of abrupt change in composition or phase change are considered reasonable regions for melting to begin. The addition of volatiles to rocks is found to be a suitable method for initiating melting at temperatures considerably below those for melting in the absence of volatiles. Perturbations in specific properties such as density or thermal conductivity may lead to the initiation of melting. Because of the decrease in thermal conductivity and viscosity with temperature, both properties provide a runaway character to some methods. Although each of these methods has its advantages, the most acceptable processes still appear to be radioactive heating and adiabatic rise.

The enthalpy of melting was found to be relatively small compared with the total amount of heat required to bring a rock up to the melting temperature. New estimates of the enthalpy of melting, when corrected for pressure, are greater than previously indicated. Millions of years are required to produce small amounts of melt even if the entire local radioactive heat production is consumed. Alternatively, the heat required for melting could be obtained from the adiabatic rise of a hot mass. It was found that the depth of initiation of the rise would have to be as great as 160 km for complete melting of eclogite, a material equivalent in composition to basalt. On the other hand, the release of liquid before complete melting would not require such great depths of initiation. It would, however, be necessary for the eclogite to melt in a eutecticlike fashion and for all fractions to be basaltic in composition. The rise of eclogite is a most unlikely event because of its high density, so a parental material such as garnet peridotite is preferred. The partial melting of garnet peridotite yields a range of basaltic liquids, the type being dependent on depth of separation, amount and proportion of volatiles, and other factors. The adiabatic rise of garnet peridotite to shallow regions with appropriate metamorphic changes, fractional melting, and separation of liquid appears to account best for the observations in the midocean ridge environment.

WHY DO BASALTIC MELTS PREDOMINATE?

Large volumes of relatively homogeneous basaltic magma have been produced throughout geologic time. The melting process must therefore yield a uniform product consistently and repetitively. The physicochemical properties of garnet peridotite outlined in the laboratory suggest that a substantial amount of liquid may be derived from it at the beginning-of-melting temperature. The liquid may be removed in one batch or fractionally as it is formed. Zone melting, which requires time-dependent, nonequilibrium processes, does not appear to be appropriate for the extended times involved in magma generation. Fractional melting is the more likely process because the local dynamic motions of the mantle readily displace a liquid from the site of origin.

Fractional melting at an invariant point, which tends to be the lowest temperature (and pressure) for melting to begin, yields a liquid of relatively uniform composition in major components but not in trace elements. The proportion of phases in the parental material does not influence the major element composition of the liquid, but it greatly influences its trace element content. Fractional melting of the parental material ceases with the loss of a phase, resuming again when the temperature reaches that of the beginning of melting of the diminished assemblage. The liquids withdrawn at each invariant point may be collected into uniform batches, differing uniquely from each other in bulk composition without intermediate compositions. It was demonstrated that the first batch generated at high pressures at the lowest temperature invariant point for the garnet peridotite system was, in a broad sense, basaltic in composition. It is for this reason that basalts are the most common magma type on earth.

WHAT FACTORS DETERMINE THE KINDS OF BASALT?

A study of the thermochemical relations of the major end-member minerals in basalts at 1 atm reveals that there are at least three invariant points involving forsterite separated by thermal maxima. These invariant points are analogous to the three major types of basalt. The eutecticlike behavior in the laboratory of natural basalts at 1 atm reflects the character implied by the analogous invariant points. The major rock types derived by crystal fractionation from basalt can be related to the univariant curves leading to lower temperatures and other subordinate invariant points. These observations demonstrate that basalts and their derivatives result from controlled physicochemical behavior.

At high pressures all basalts convert to eclogite, and a new set of phase relations obtains. These relations have not yet been determined in detail; present evidence suggests, however, that a single parental assemblage could yield the various liquids of basaltic composition with suitable changes in pressure or volatile composition. In general, high pressure and CO_2-rich volatiles produce alkali basalts, whereas low pressure and H_2O-rich volatiles produce tholeiitic varieties. The variation in trace elements may be attributed to different proportions of phases in the parental assemblage, the degree of melting, and the presence of various minor phases. It is essential to determine the depth of separation of the liquid from the parental material and the amount and ratios of the volatile constituents. Other factors, such as the partial pressure of oxygen, also influence the nature of the liquid derived from the partial or fractional melting of garnet peridotite.

The absence of olivine in eclogites produced at high pressures from basalts would appear to cast doubt on the derivation of basaltic liquids from an olivine-bearing parental material such as garnet peridotite. The absence of olivine can be accounted for either by the removal of phases from the liquid during its ascent to the surface or by the existence of a reaction relation involving olivine and liquid. An additional reaction relation involving orthopyroxene and liquid is necessary to account for the production of an eclogitic liquid from garnet peridotite. Both reaction relations have been demonstrated in the laboratory. The sequence of reactions is believed to involve the removal first of olivine and then of orthopyroxene, with decreasing pressure; however, the exact compositional trend of the liquids as a result of these reactions at high pressures in the absence of volatiles has not yet been established. Removal of olivine tends to increase the normative quartz content of the liquid, whereas removal of orthopyroxene tends to increase the normative nepheline or larnite content of the liquid. There is the prospect of generating larnite-normative liquids (potential melilite nephelinites at low pressures) if, with decreasing pressure, the orthopyroxene reacts out first.

The proportions of garnet and clinopyroxene in garnet peridotite are a measure of the degree of depletion of the parental material of its basaltic fraction. From the viewpoint of partial or fractional melting, the order in which clinopyroxene or garnet is consumed in the liquid influences the kind of basaltic liquid produced. Phase-equilibrium studies suggest that clinopyroxene may be the first phase consumed in the partial melting of garnet peridotite at high pressures, whereas garnet is consumed first at moderate pressures for compositions appropriate to basalts. The norms of natural garnets and clinopyroxenes from garnet peridotites clearly illustrate the great contrast in chemical contribution of these phases to the melt.

HOW DOES THE MELT FORM AND AGGREGATE?

The first melt most likely appears at the common junction of all phases. Because all phases are in equilibrium with the initial liquid, the liquid is not confined to the initial pore. The liquid is presumed to wet all crystals and permeates all grain boundaries, forming a "liquid-bonded aggregate." Furthermore, the concentration of impurities and the low-temperature melting residua occurring at the grain boundaries of the major phases facilitate the melting process. A significant drop in the velocity of transmission of seismic waves occurs at this stage. The aggregation is aided by the thermal gradient and shear stresses on the host rock. In general, the liquid aggregates preferably in regions perpendicular to the principal minimum stress and may be kneaded out under very low stresses.

The region of magma generation is essentially defined by the solidus of the rock undergoing melting. Because the solidus varies with pressure, the boundary will not necessarily be an isotherm. The nature of the heat source of the magma, whether external heat transfer or internal heat production, will determine, to a large extent, the initial size and shape of the melting region. The boundary, sharply defined by the appearance of liquid, may be diffuse in regard to some physical properties, yet sharp in its response to seismic wave transmission. A plastic envelope no doubt surrounds the chamber as a result of heat transfer.

Melting takes place with a large increase in volume. Space must be provided for the volume increase by tumescence of the region, filling of pore spaces in adjoining host rock, or rupture of the host rock. If space is not provided, the magma will freeze because the melting temperature generally increases with pressure. The presence of melt lowers the strength of the containing rock, and flaws in the plastic envelope are expanded by a magmafracting process. Earthquakes may accompany the magmafracting process as the magma enters the brittle region beyond the plastic envelope.

Differential stress tends to raise the mean stress on a volume of rock. The melt does not freeze under the small increase in pressure because (a) the ratio of compressibility to thermal expansion is favorable, (b) the low thermal conductivity of the host slows the rate of release of the enthalpy of melting, or (c) the presence of volatiles may enhance the melting under an increase in pressure.

WHAT ARE THE FORCES CAUSING MAGMA RELEASE?

The main force causing magma release probably results from the pressure of the rock overburden. The loss of strength of the partially molten rock initiates the magmafracting process, and the liquid first migrates through the plastic envelope and then propagates fractures to the surface. The density difference between the containing rock and the anastomosing magma column results in a pressure difference. The gravitational force of the rock column on the magma, exceeding that of the magma column itself, causes the magma to rise.

If an appropriate viscosity of the mantle is assumed, the rate of rise of small (l-km-diameter) liquid spheres by buoyancy could account for the present worldwide production. Because heat losses from the magma en route to the surface and extensive lateral penetration of the crust by sills terminate many potential extrusions, large liquid masses must be periodically released from the zone of partial or fractional melting. Depending on the rate of rise, the masses may remain intact in batholithic proportions or disintegrate into a series of smaller masses.

Overburden pressure and buoyancy appear to be adequate forces to account for the rise of magma; however, they are probably not the primary forces driving the large-scale convection in the mantle envisaged by some investigators. Tidal forces may trigger volcanic action, but the major disruptions that initiate the worldwide geological revolutions have not been identified.

WHY ARE VOLCANIC ERUPTIONS PERIODIC?

Since the beginning of the earth's history, there have been at least seven major periods of igneous activity based on mineral "age." Individual volcanoes are themselves periodic and appear to have a finite lifetime of the order of 1–5 m.y. The great floods of basalt have appeared in a relatively short time-span of a few to 20 m.y. The accurate dating of older basalts is yet to be perfected, and many more measurements are needed before major conclusions can be drawn.

The Joly–Cotter mechanism of convective heat transfer, unsatisfactory in several details, appears to provide the only available long-time-scale process for the igneous activity associated with geological revolutions. The short-period eruptions of individual volcanoes may be closely related to the continuous filling of a subsidiary reservoir that swells and breaks periodically. The thermal effect on some properties, such as viscosity and thermal conductivity, leads to runaway conditions that may be periodic as long as the energy source persists. The recovery time may be long because of the slow rate of heating and the relatively large enthalpy of melting. Recurring explosive eruptions may be related to the volatile saturation of a magma as a result of cooling or pressure release.

The mechanisms of magma generation have, no doubt, changed with geologic time. The early crust, no remnant of which has been identified, was probably andesitic because of the hydrous melting attending the degassing of the earth to form the oceans. The more recent increase in alkali basalt generation during the last 250 m.y. (Engel *et al.*, 1965, p. 729) also suggests a style of magma generation different from that producing the midocean upwelling of magma now considered essential to plate tectonic theory. Poldervaart (1962, p. 10) was unable to find any record of alkali basalts older than 1 b.y. and concluded that tholeiite is therefore the "primitive" magma. The decline in production of tholeiitic magmas may be reflecting the general decrease in anticipated heat production due to decay of radioactive elements.

CONCLUSION

The exciting questions now amenable to experiment ensure an expanding effort in the study of magma generation. There is much for the physicist, thermodynamicist, and physical chemist to ponder, provided the questions are framed within the constraints deduced by the field geologist, petrologist, geophysicist, and mineralogist. Recognition of the relevant parameters is indeed a critical contribution to the problem. Skepticism of the model approach to earth problems is warranted because many key parameters have not been included. One can readily sense a new revolution in geologic thought coming close on the heels of the plate-tectonic awakening.

Appendix: Conversion Factors

ENERGY

1 joule $= 10^7$ erg
1 erg $= 0.23901 \times 10^{-7}$ cal
1 cal $= 4.1840 \times 10^7$ erg
1 erg $= 1$ g cm^2/sec^2

FORCE

1 dyne $= 1$ g cm/sec^2

GRAVIMETRIC FACTORS

$Fe_2O_3 \times 0.89981 = FeO$
$FeO \times 1.1113 = Fe_2O_3$
$FeO \times 0.77731 = Fe$
$Fe_2O_3 \times 0.69944 = Fe$
$K_2O \times 0.83015 = K$

GRAVITATIONAL ACCELERATION

Normal $g_{0°} = 978.032$ cm/sec^2
1 Gal $= 1$ cm/sec^2

LENGTH

1 km $= 10^5$ cm

PRESSURE

1 bar $= 10^6$ dyne/cm^2
1 bar $= 10^6$ g/cm/sec^2
1 mm Hg $= 1$ torr $= 1.33322 \times 10^{-3}$ bar

TIME

1 yr $= 3.1536 \times 10^7$ sec

VELOCITY

1 km/yr $= 3.171 \times 10^{-2}$ cm/sec

VISCOSITY

1 poise $= 1$ g/cm/sec

RARE-EARTH NORMALIZATION

Average rare-earth element composition in parts per million of chondrites used in normalizing abundances of those elements in rocks and minerals

Element	Schmitt et al. (1963, 1964) (average of 20)	Haskin et al. (1968) (average of 9)
La	0.30 ± 0.06	0.330 ± 0.013
Ce	0.84 ± 0.18	0.88 ± 0.01
Pr	0.12 ± 0.02	0.112 ± 0.005
Nd	0.58 ± 0.13	0.60 ± 0.01
Sm	0.21 ± 0.04	0.181 ± 0.006
Pm	—	—
Eu	0.074 ± 0.015	0.069 ± 0.001
Gd	0.32 ± 0.07	0.249 ± 0.011
Tb	0.049 ± 0.010	0.047 ± 0.001
Dy	0.31 ± 0.019	—
Ho	0.073 ± 0.014	0.070 ± 0.001
Er	0.21 ± 0.04	0.200 ± 0.005
Tm	0.033 ± 0.007	0.030 ± 0.002
Yb	0.17 ± 0.03	0.200 ± 0.007
Lu	0.031 ± 0.005	0.034 ± 0.002

References

Adams, L. H. (1922) Temperature changes accompanying isentropic, isenergic, and isenkaumic expansion, *J. Wash. Acad. Sci., 12,* 407–411.

Adams, L. H., and R. E. Gibson (1926) The compressibilities of dunite and of basalt glass and their bearing on the composition of the earth, *Proc. Natl. Acad. Sci. U.S.A., 12,* 275–283.

Adams, L. H., and E. D. Williamson (1923) The compressibility of minerals and rocks at high pressures, *J. Franklin Inst., 195,* 475–529.

Adams, L. H., and E. D. Williamson (1925) The composition of the earth's interior, *Smithson. Rep. 1923,* 241–260.

Akella, J. (1974) Solubility of Al_2O_3 in orthopyroxene coexisting with garnet and clinopyroxene for compositions on the diopside-pyrope join in the system $CaSiO_3$–$MgSiO_3$–Al_2O_3, *Carnegie Inst. Washington Yearb. 73,* 273-278.

Akella, J., and F. R. Boyd (1974) Petrogenetic grid for garnet peridotites, *Carnegie Inst. Washington Yearb. 73,* 269–273.

Akella, J., and G. C. Kennedy (1971) Studies on anorthite + diopside$_{50}$–hedenbergite$_{50}$ at high pressures and temperatures, *Am. J. Sci., 270,* 155–165.

Akimoto, S.-I. (1972) The system MgO–FeO–SiO_2 at high pressures and temperatures—phase equilibria and elastic properties. In A. R. Risema, ed., *The upper mantle,* Elsevier Publishing Company, New York, p. 161–187.

Allen, E. T., and W. P. White (1909) Diopside and its relation to calcium and magnesium metasilicates, *Am. J. Sci., 27,* 1–47.

Allison, E. B., P. Brock, and J. White (1959) The rheology of aggregates containing a liquid phase with special reference to the mechanical properties of refractories at high temperatures, *Trans. Br. Ceram. Soc., 58,* 495–531.

Anderson, D. L. (1962) The plastic layer of the earth's mantle, *Sci. Am., 207,* 52–59.

Anderson, D. L. (1970) Petrology of the mantle, *Mineral. Soc. Am. Spec. Pap. 3,* 85–93.

Anderson, D. L., and H. Spetzler (1970) Partial melting and the low-velocity zone, *Phys. Earth Planet. Inter., 4,* 62–64.

Anderson, O. (1915) The system anorthite–forsterite–silica, *Am. J. Sci., 39,* 407–454.

Anderson, O. L., and P. C. Perkins (1975) A plate tectonics model involving non-laminar asthenospheric flow to account for irregular patterns of magmatism in the southwestern United States, *Phys. Chem. Earth, 9*, 113–122.

Anthony, T. R., and H. E. Cline (1970) The kinetics of droplet migration in solids in an accelerational field, *Philos. Mag., 22*, 893–901.

Anthony, T. R., and H. E. Cline (1971) Thermal migration of liquid droplets through solids, *J. Appl. Phys., 42*, 3380–3387.

Anthony, T. R., and H. E. Cline (1972a) The thermomigration of biphase vapor–liquid droplets in solids, *Acta Metall., 20*, 247-255.

Anthony, T. R., and H. E. Cline (1972b) The interaction of migrating liquid inclusions with grain boundaries in solids. In Hsun Hu, ed., *Nature and behavior of grain boundaries*, Plenum Publishing Corp., New York, p. 185–201.

Arculus, R. J. (1975) Melting behavior of two basanites in the range 10 to 35 kbar and the effect of TiO₂ on the olivine–diopside reactions at high pressures, *Carnegie Inst. Washington Yearb. 74*, 512–515.

Arculus, R. J., and E. B. Curran (1972) The genesis of the calc-alkaline rock suite, *Earth Planet. Sci. Lett., 15*, 255–262.

Arculus, R. J., and N. Shimizu (1974) Rare earth elements in a suite of basanitoids and alkali olivine basalts from Grenada, Lesser Antilles, *Carnegie Inst. Washington Yearb. 73*, 553–560.

Ashby, M. F. (1972) A first report on deformation-mechanism maps, *Acta Metall., 20*, 887–897.

Bailey, D. K. (1970) Volatile flux, heat-focusing and the generation of magma. In G. Newall and N. Rast, eds., *Mechanism of igneous intrusion*, Gallery Press, Liverpool, p. 177–186.

Barton, D. C. (1936) Mechanics of formation of salt domes with special reference to Gulf Coast salt domes of Texas and Louisiana. In D. C. Barton and G. Sawtelle, eds., *Gulf Coast Oil Fields; a Symposium on the Gulf Coast Cenozoic*, American Association of Petroleum Geologists, Tulsa, Oklahoma, p. 20–78.

Barus, C. (1893) High-temperature work in igneous fusion and ebullition, chiefly in relation to pressure, *U.S. Geol. Surv. Bull. 103*, 57 p.

Bath, M. (1966) Earthquake energy and magnitude, *Phys. Chem. Earth, 7*, 115–165.

Bayly, B. (1968) *Introduction to petrology*, Prentice-Hall, Inc., Englewood Cliffs, New Jersey, 371 p.

Beeson, M. H., and E. D. Jackson (1970) Origin of the garnet pyroxenite xenoliths at Salt Lake Crater, Oahu, *Mineral. Soc. Am. Spec. Pap. 3*, 95–112.

Bell, P. M., and H. K. Mao (1972) The hypothesis of melting at stress dislocations in the earth, *Carnegie Inst. Washington Yearb. 71*, 416–418.

Bell, P. M., and H. K. Mao (1975) Laser optical system for heating experiments and pressure calibration of the diamond-windowed, high-pressure cell, *Carnegie Inst. Washington Yearb. 74*, 399–402.

Benioff, H. (1954) Orogenesis and deep crustal structure—additional evidence from seismology, *Geol. Soc. Am. Bull., 65*, 385–400.

Betz, F., Jr., and H. H. Hess (1942) The floor of the North Pacific Ocean, *Geogr. Rev., 32*, 99–116.

Biggs, W. D. (1970) Fracture. In R. W. Cann, ed., *Physical metallurgy*, North-Holland Publishing Company, Amsterdam, p. 1199–1232.

Birch, F. (1951) Recent work on the radioactivity of potassium and some related geophysical problems, *J. Geophys. Res., 56*, 107–126.

Birch, F. (1964) Mega-geological considerations in rock mechanics. In W. R. Judd, ed., *State of stress in the earth's crust*, American Elsevier Publishing Company, Inc., New York, p. 55–80.

Birch, F. (1965) Speculations on the earth's thermal history, *Geol. Soc. Am. Bull.*, *76*, 133–153.

Birch, F. (1969) Density and composition of the upper mantle: first approximation as an olivine layer. In P. J. Hart, ed., The earth's crust and upper mantle, *Geophys. Monogr., Am. Geophys. Union*, *13*, 18–36.

Birch, F. (1970) Interpretations of the low-velocity zone, *Phys. Earth Planet. Inter.*, *3*, 178–181.

Birch, F., and D. Bancroft (1942) The elasticity of glass at high temperatures, and the vitreous basaltic substratum, *Am. J. Sci.*, *240*, 457–490.

Birch, F., and H. Clark (1940) The thermal conductivity of rocks and its dependence upon temperature and composition, *Am. J. Sci.*, *238*, 529–558.

Birch, F., and P. LeComte (1960) Temperature-pressure plane for albite composition, *Am. J. Sci.*, *258*, 209–217.

Boettcher, A. L., and P. J. Wyllie (1969) Phase relationships in the system $NaAlSiO_4$–SiO_2–H_2O to 35 kilobars pressure, *Am. J. Sci.*, *267*, 875–909.

Bowden, F. P., M. A. Stone, and G. K. Tudor (1947) Hot spots on rubbing surfaces and the detonation of explosives by friction, *Proc. R. Soc. London, Ser. A, 188,* 329–349.

Bowden, F. P., and P. H. Thomas (1954) The surface temperature of sliding solids, *Proc. R. Soc. London, Ser. A, 223,* 29–40.

Bowen, N. L. (1915) The crystallization of haplobasaltic, haplodioritic, and related magmas, *Am. J. Sci., 40,* 161–185.

Bowen, N. L. (1928) *The evolution of the igneous rocks*, Princeton University Press, Princeton, New Jersey, 332 p.

Bowen, N. L., and J. F. Schairer (1935) The system MgO–FeO–SiO_2, *Am. J. Sci., 29,* 151–217.

Boyd, F. R. (1970) Garnet peridotites and the system $MgSiO_3$–$CaSiO_3$–Al_2O_3, *Mineral. Soc. Am. Spec. Pap. 3,* 63–75.

Boyd, F. R., and J. L. England (1959a) Pyrope, *Carnegie Inst. Washington Yearb. 58,* 83–87.

Boyd, F. R., and J. L. England (1959b) Quartz–coesite transition, *Carnegie Inst. Washington Yearb. 58,* 87–88.

Boyd, F. R., and J. L. England (1963) Effect of pressure on the melting of diopside, $CaMgSi_2O_6$, and albite, $NaAlSi_3O_8$, in the range up to 50 kilobars, *J. Geophys. Res., 68,* 311–323.

Boyd, F. R., J. L. England, and B. T. C. Davis (1964) Effects of pressure on the melting and polymorphism of enstatite, $MgSiO_3$, *J. Geophys. Res., 69,* 2101–2109.

Boyd, F. R., and P. H. Nixon (1973) Structure of the upper mantle beneath Lesotho, *Carnegie Inst. Washington Yearb. 72,* 431–445.

Boyd, F. R., and P. H. Nixon (1975) Origins of the ultramafic nodules from some kimberlites of northern Lesotho and the Monastery Mine, South Africa, *Phys. Chem. Earth, 9,* 431–454.

Brace, W. F. (1964) Brittle fracture of rocks. In W. R. Judd, ed., *State of stress in the earth's crust*, American Elsevier Publishing Company, Inc., New York, p. 111–178.

Brace, W. F., W. C. Luth, and J. D. Unger (1968) Melting of granite under an effective confining pressure (abstract), *Geol. Soc. Am. Spec. Pap. 115,* 21.

Bradley, R. S. (1962) Thermodynamic calculations on phase equilibria involving fused salts, Part II, Solid solutions and application to the olivines, *Am. J. Sci., 260,* 550–554.

Bridgman, P. W. (1948) Rough compressions of 177 substances to 40,000 kg/cm², *Proc. Am. Acad. Arts Sci., 76,* 71–87.

Brooks, C., D. E. James, and J. R. Hart (1976) Ancient lithosphere: Its role in young continental volcanism, *Science 193,* 1086–1094.

Brune, J. N., and C. R. Allen (1967) A low-stress-drop, low magnitude earthquake with surface faulting: the Imperial, California, earthquake of March 4, 1966, *Bull. Seismol. Soc. Am., 57,* 501–514.

Buddington, A. F. (1943) Some petrological concepts and the interior of the earth, *Am. Mineral., 28,* 119–140.

Bullen, K. E. (1947) *An introduction to the theory of seismology,* Cambridge University Press, Cambridge, 276 p.

Bultitude, R. J., and D. H. Green (1971) Experimental study of crystal–liquid relationship at high pressures in olivine nephelinite and basanite compositions, *J. Petrol., 12,* 121–147.

Burnham, C. W., and N. F. Davis (1974) The role of H_2O in silicate melts, II, Thermodynamic and phase relations in the system $NaAlSi_3O_8–H_2O$ to 10 kilobars, 700° to 1100°C, *Am. J. Sci., 274,* 902–940.

Butkovich, T. R. (1974) *Rock melt from an underground nuclear explosion,* Lawrence Livermore Laboratory, UCRL-51554, Technical Information Document 4500, UC-35, National Technical Information Service, Springfield, Virginia.

Butler, B. C. M. (1961) Metamorphism and metasomatism of rocks of the Moine series by a dolerite plug in Glenmore, Ardnamurchan, *Mineral. Mag., 32,* 866–897.

Cameron, M., S. Sueno, C. T. Prewitt, and J. J. Papike (1973) High-temperature crystal chemistry of acmite, diopside, hedenbergite, jadeite, spodumene, and ureyite, *Am. Mineral., 58,* 594–618.

Carr, M. J., and R. E. Stoiber (1973) Intermediate depth earthquakes and volcanic eruptions in Central America, 1961–1972, *Bull. Volcanol., 37,* 326–337.

Carter, N. L., and H. G. Ave'Lallemant (1970) High temperature flow of dunite and peridotite, *Geol. Soc. Am. Bull., 81,* 2181–2202.

Cawthorn, R. G. (1975) Degrees of melting in mantle diapirs and the origin of ultrabasic liquids, *Earth Planet. Sci. Lett., 27,* 113–120.

Chamberlin, T. C., and R. D. Salisbury (1905) *Geology, Vol. II, Earth history,* 2d ed., Henry Holt & Co., New York, 692 p.

Chamberlin, T. C., and R. D. Salisbury (1909) *Geology, Vol. I, Geologic processes and their results,* 2d ed., Henry Holt & Co., New York, 684 p.

Chayes, F. (1963) Relative abundance of intermediate members of the oceanic basalt–trachyte association, *J. Geophys. Res., 68,* 1519–1534.

Chayes, F. (1972) Silica saturation in Cenozoic basalt, *Philos. Trans. R. Soc. London, Ser. A, 271,* 285–296.

Chayes, F. (1975) Average composition of the commoner Cenozoic volcanic rocks, *Carnegie Inst. Washington Yearb. 74,* 547–549.

Chen, C.-H., and D. C. Presnall (1975) The system $Mg_2SiO_4–SiO_2$ at pressures up to 25 kilobars, *Am. Mineral., 60,* 398–406.

Clark, P. W., J. H. Cannon, and J. White (1953) Further investigations on the sintering of oxides, *Trans. Br. Ceram. Soc., 52,* 1–49.

Clark, S. P., Jr., and A. E. Ringwood (1964) Density distribution and constitution of the mantle, *Rev. Geophys., 2,* 35–88.

Clark, S. P., Jr., J. F. Schairer, and J. de Neufville (1962) Phase relations in the system $CaMgSi_2O_6–CaAl_2SiO_6–SiO_2$ at low and high pressure, *Carnegie Inst. Washington Yearb. 61,* 59–68.

Clarke, W. B., M. A. Beg, and H. Craig (1969) Excess ^{3}He in the sea: evidence for terrestrial primordial helium, *Earth Planet. Sci. Lett., 6,* 213–220.

Clayton, R. N., and T. K. Mayeda (1975) Genetic relations between the moon and meteorites, *Proc. Sixth Lunar Sci. Conf., Geochim. Cosmochim. Acta (Suppl. 6), 2,* 1761–1769.

Coats, R. R. (1962) Magma type and crustal structure in the Aleutian arc. In G. A. Macdonald and H. Kuno, eds., The crust of the Pacific basin, *Geophys. Monogr., Am. Geophys. Union, 6,* 92–109.

Coble, R. L., and J. E. Burke (1963) Sintering in ceramics. In J. E. Burke, ed., *Progress in ceramic science,* Macmillan Company, New York, p. 197–251.

Coe, R. S., and M. S. Paterson (1969) The α–β inversion in quartz: a coherent phase transition under nonhydrostatic stress, *J. Geophys. Res., 74,* 4921–4928.

Cohen, L. H., K. Ito, and G. C. Kennedy (1967) Melting and phase relations in an anhydrous basalt to 40 kilobars, *Am. J. Sci., 265,* 475–518.

Cordani, U. G., and P. Vandoros (1967) Basaltic rocks of the Paraná Basin. In J. J. Bigarella, R. D. Becker, and I. D. Pinto, eds., *Problems in Brazilian Gondwana geology,* International Symposium on the Gondwana Stratigraphy and Palaeontology, 1st, Argentina, 1967, Curitiba, Brazil, p. 207–231.

Cotter, J. R. (1924) The escape of heat from the earth's crust, *Philos. Mag., 48,* 458–464.

Cox, K. G. (1970) Tectonics and vulcanism of the Karroo Period and their bearing on the postulated fragmentation of Gondwanaland. In T. N. Clifford and I. G. Gass, eds., *African magmatism and tectonics,* Hafner Publishing Company, Darien, Connecticut, p. 211–235.

Crandell, D. R., D. R. Mullineaux, and M. Rubin (1975) Mount St. Helens volcano: recent and future behavior, *Science, 187,* 438–441.

Crittenden, M. D., Jr. (1967) Viscosity and finite strength of the mantle as determined from water and ice loads, *Geophys. J. R. Astron. Soc., 14,* 261–279.

Cullers, R. L., L. G. Medaris, Jr., and L. A. Haskin (1970) Gadolinium: distribution between aqueous and silicate phases, *Science, 169,* 580–583.

Daly, R. A. (1911) The nature of volcanic action, *Proc. Am. Acad. Arts Sci., 47,* 48–122.

Daly, R. A. (1925a) Relation of mountain building to igneous action, *Proc. Am. Philos. Soc., 64,* 283–307.

Daly, R. A. (1925b) The geology of Ascension Island, *Proc. Am. Acad. Sci., 60,* 1–80.

Daly, R. A. (1933) *Igneous rocks and the depths of the earth,* 2d ed., McGraw-Hill Book Company, New York, 598 p.

Davis, B. T. C. (1964) The system diopside–forsterite–pyrope at 40 kilobars, *Carnegie Inst. Washington Yearb. 63,* 165–171.

Davis, B. T. C., and F. R. Boyd (1966) The join $Mg_2Si_2O_6$–$CaMgSi_2O_6$ at 30 kilobars pressure and its application to pyroxenes from kimberlites, *J. Geophys. Res., 71,* 3567–3576.

Davis, B. T. C., and J. L. England (1964) The melting of forsterite up to 50 kilobars, *J. Geophys. Res., 69,* 1113–1116.

Davis, B. T. C., and J. F. Schairer (1965) Melting relations in the join diopside–forsterite–pyrope at 40 kilobars and at one atmosphere, *Carnegie Inst. Washington Yearb. 64,* 123–126.

Davis, N. F. (1972) Experimental studies in the system $NaAlSi_3O_8$–H_2O, Part I, The apparent solubility of albite in supercritical water; Part II, The partial specific volume of H_2O in $NaAlSi_3O_8$ melts with petrologic implications, unpublished Ph.D. thesis, Pennsylvania State University, University Park, Pennsylvania.

Day, A. L., and E. S. Shepherd (1913) Water and volcanic activity, *Geol. Soc. Am. Bull., 24*, 573–606.

DeLury, J. S. (1944) Generation of magma by frictional heat, *Am. J. Sci., 242*, 113–129.

de Wys, E. C., and W. R. Foster (1958) The system diopside–anorthite–åkermanite, *Mineral. Mag., 31*, 736–743.

Dickey, J. S., Jr. (1970) Partial fusion products in Alpine-type peridotites: Serrania de la Ronda and other examples, *Mineral. Soc. Am. Spec. Pap. 3*, 33–49.

Dickey, J. S., Jr., and H. S. Yoder, Jr. (1972) Partitioning of chromium and aluminum between clinopyroxene and spinel, *Carnegie Inst. Washington Yearb. 71*, 384–392.

Dubourdieu, G. G. (1973) On the four-year seismic period, private publication, translated from the French by Noël Lindsay.

Dunn, P. R., and M. C. Brown (1969) North Australian plateau volcanics, *Spec. Publ. Geol. Soc. Aust., 2*, 117–122.

Du Toit, A. (1937) *Our wandering continents*, Oliver and Boyd, Edinburgh, 366 p.

Eaton, J. P. (1959) A portable water-tube tiltmeter, *Bull. Seismol. Soc. Am., 49*, 301–316.

Eaton, J. P. (1962) Crustal structure and volcanism in Hawaii. In G. A. Macdonald and H. Kuno, eds., The crust of the Pacific basin, *Geophys. Monogr., Am. Geophys. Union, 6*, 13–29.

Eaton, J. P., R. L. Christiansen, H. M. Iyer, A. M. Pitt, D. R. Mabey, H. R. Blank, Jr., I. Zietz, and M. E. Gettings (1975) Magma beneath Yellowstone National Park, *Science, 188*, 787–796.

Eaton, J. P., and K. J. Murata (1960) How volcanoes grow, *Science, 132*, 925–938.

Eggler, D. H. (1973) Role of CO_2 in melting processes in the mantle, *Carnegie Inst. Washington Yearb. 72*, 457–467.

Eggler, D. H. (1974) Effect of CO_2 on the melting of peridotites, *Carnegie Inst. Washington Yearb. 73*, 215–224.

Eggler, D. H. (1975) Peridotite–carbonatite relations in the system $CaO–MgO–SiO_2–CO_2$, *Carnegie Inst. Washington Yearb. 74*, 468–474.

Eggler, D. H., B. O. Mysen, and M. G. Seitz (1974) The solubility of CO_2 in silicate liquids and crystals, *Carnegie Inst. Washington Yearb. 73*, 226–228.

Einstein, A. (1906) Eine neue Bestimmung der Moleküldimensionen, *Ann. Phys. (Leipzig), 19*, 289–306.

Einstein, A. (1911) Berichtigung zu meiner Arbeit: "Eine neue Bestimmung der Moleküldimensionen," *Ann. Phys. (Leipzig), 34*, 591–592.

Emslie, R. F. (1971) Liquidus relations and subsolidus reactions in some plagioclase-bearing systems, *Carnegie Inst. Washington Yearb. 69*, 148–155.

Emslie, R. F., and D. H. Lindsley (1969) Experiments bearing on the origin of anorthositic intrusions, *Carnegie Inst. Washington Yearb. 67*, 108–109.

Engel, A. E. J., and C. G. Engel (1964) Composition of basalts from the mid-Atlantic Ridge, *Science, 144*, 1330–1333.

Engel, A. E. J., C. G. Engel, and R. G. Havens (1965) Chemical characteristics of oceanic basalts and the upper mantle, *Geol. Soc. Am. Bull., 76*, 719–733.

Erlank, A. J., and E. J. D. Kable (1976) The significance of incompatible elements in Mid-Atlantic Ridge basalts from 45° N with particular reference to Zr/Nb, *Contrib. Mineral. Petrol., 54*, 281–291.

Eshelby, V. D. (1957) The determination of the elastic field of an ellipsoidal inclusion, and related problems, *Proc. R. Soc. London, Ser. A, 241*, 376–396.

Evans, A. G., M. Linzer, and L. R. Russel (1974) Acoustic emission and crack propagation in polycrystalline alumina, *Mater. Sci. Eng., 15*, 253–261.

Fahrig, W. F., and R. K. Wanless (1963) Age and significance of diabase dyke swarms of the Canadian shield, *Nature (London), 200,* 934–937.

Faure, G., and J. L. Powell (1972) *Strontium isotope geology,* Springer-Verlag, New York, 188 p.

Fedotov, S. A., and P. I. Tokarev (1974) Earthquakes, characteristics of the upper mantle under Kamchatka and their connection with volcanism (according to data collected up to 1971), *Bull. Volcanol., 37,* 245–257.

Feild, A. L., and P. H. Royster (1918) Slag viscosity tables for blast-furnace work, *U.S. Bur. Mines Tech. Pap. 187,* 38 p.

Ferguson, J. B., and H. E. Merwin (1919) The ternary system $CaO-MgO-SiO_2$, *Am. J. Sci., 48,* 81–123.

Fermor, L. L. (1913) Preliminary note on garnet as a geological barometer and on an infra-plutonic zone in the earth's crust, *Rec. Geol. Surv. India, 43,* Pt. 1, 41–47.

Finger, L. W., and Y. Ohashi (1976) The thermal expansion of diopside to 800°C and a refinement of the crystal structure at 700°C, *Am. Mineral., 61,* 303–310.

Fiske, R. S., C. A. Hopson, and A. C. Waters (1963) Geology of Mount Rainier National Park, Washington, *U.S. Geol. Surv. Prof. Pap. 444,* 93 p.

Fiske, R. S., and E. D. Jackson (1972) Orientation and growth of Hawaiian volcanic rifts: the effect of regional structure and gravitational stresses, *Proc. R. Soc. London, Ser. A, 329,* 299–326.

Foster, W. R. (1942) The system $NaAlSi_3O_8-CaSiO_3-NaAlSiO_4$, *J. Geol., 50,* 152–173.

Fujii, T., and E. Takahashi (1976) On the solubility of alumina in orthopyroxene coexisting with olivine and spinel in the system $MgO-Al_2O_3-SiO_2$, *Mineral. J., 8,* No. 2, 122–128.

Gast, P. W. (1960) Limitations on the composition of the upper mantle, *J. Geophys. Res., 65,* 1287–1297.

Gast, P. W. (1968) Trace element fractionation and the origin of tholeiitic and alkaline magma types, *Geochim. Cosmochim. Acta, 32,* 1057–1086.

Gast, P. W., G. R. Tilton, and C. Hedge (1964) Isotopic composition of lead and strontium from Ascension and Gough Islands, *Science, 145,* 1181–1185.

Gastil, G. (1960) The distribution of mineral dates in time and space, *Am. J. Sci., 258,* 1–35.

Gault, D. E., and E. D. Heitowit (1963) The partition of energy for hypervelocity impact craters formed in rock, *Proc. Sixth Hypervelocity Impact Symp.,* Cleveland, Ohio, *2,* 419–456.

Geological Museum of London (1972) *The story of the earth,* Institute of Geological Sciences, Geological Museum, London, Her Majesty's Stationery Office, 36 p.

Gilbert, M. C. (1969) Reconnaissance study of the stability of amphiboles at high pressure, *Carnegie Inst. Washington Yearb. 67,* 167–170.

Gilluly, J. (1972) Tectonics involved in the evolution of mountain ranges. In E. C. Robertson, ed., *Nature of the solid earth,* McGraw-Hill Book Company, New York, p. 406–439.

Goetze, C. (1975) Sheared lherzolites: from the point of view of rock mechanics, *Geology, 3,* 172–173.

Gorshkov, G. S. (1958) On some theoretical problems of volcanology, *Bull. Volcanol., 19,* 103–113.

Green, D. H. (1971) Composition of basaltic magmas as indicators of conditions of origin: application to oceanic volcanism, *Philos. Trans. R. Soc. London, Ser. A, 268,* 707–725.

Green, D. H. (1973) Contrasted melting relations in a pyrolite upper mantle under mid-oceanic ridge, stable crust and island arc environments, *Tectonophysics, 17,* 285–297.

Green, D. H., and W. Hibberson (1970) The instability of plagioclase in peridotite at high pressure, *Lithos, 3,* 209–221.

Green, D. H., I. A. Nicholls, M. Viljoen, and R. Viljoen (1975) Experimental demonstration of the existence of peridotitic liquids in earliest Archean magmatism, *Geology, 3,* 11–14.

Green, D. H., and A. E. Ringwood (1967) The genesis of basaltic magmas, *Contrib. Mineral. Petrol., 15,* 103–190.

Green, W. L. (1887) *Vestiges of the molten globe, Part II, The earth's surface features and volcanic phenomena,* Hawaiian Gazette Publishing Company, Hawaii, 337 p.

Griggs, D. T. (1954) High pressure phenomena with applications to geophysics. In L. H. Ridenour, ed., *Modern physics for the engineer,* McGraw-Hill Book Company, New York, 499 p.

Griggs, D. T., and D. W. Baker (1969) The origin of deep-focus earthquakes. In H. Mark and S. Fernbach, eds., *Properties of matter under unusual conditions,* Interscience Publishers, New York, p. 23–42.

Griggs, D., and J. Handin (1960) Observations on fracture and a hypothesis of earthquakes. In D. Griggs and J. Handin, eds., Rock deformation, *Geol. Soc. Am. Mem. 79,* 347–364.

Grout, F. F. (1945) Scale models of structures related to batholiths, *Am. J. Sci., Daly Vol., 243A,* 260–284.

Gruntfest, I. J. (1963) Thermal feedback in liquid flow; plane shear at constant stress, *Trans. Soc. Rheol., 7,* 195–207.

Gummer, W. K. (1943) The system $CaSiO_3–CaAl_2Si_2O_8–NaAlSiO_4$, *J. Geol., 51,* 503–530.

Gutenberg, B. (1926) Untersuchungen zur Frage, bis zu Welcher tiefe die erde Kristallin ist, *Z. Geophys., 2,* 24–29.

Gutenberg, B. (1959) *Physics of the earth's interior,* Academic Press, New York, 240 p.

Gutenberg, B., and C. F. Richter (1949) *Seismicity of the earth and associated phenomena,* Princeton University Press, Princeton, New Jersey, 273 p.

Gutenberg, B., and C. F. Richter (1956) Magnitude and energy of earthquakes, *Ann. Geofis., 9,* 1–15.

Haller, W. (1965) Rearrangement kinetics of the liquid–liquid immiscible microphases in alkali borosilicate melts, *J. Chem. Phys., 42,* 686–693.

Handin, J., R. V. Hager, Jr., M. Friedman, and J. N. Feather (1963) Experimental deformation of sedimentary rocks under confining pressure: pore pressure tests, *Bull. Am. Assoc. Pet. Geol., 47,* 717–755.

Harker, A. (1909) *The natural history of igneous rocks,* Methuen & Company, London, 384 p.

Harris, P. G. (1957) Zone refining and the origin of potassic basalts, *Geochim. Cosmochim. Acta, 12,* 195–208.

Hart, S. R. (1971) K, Rb, Cs, Sr, and Ba contents and Sr isotope ratios of ocean floor basalts, *Philos. Trans. R. Soc. London, Ser. A, 268,* 573–587.

Hart, S. R., J. G. Schilling, and J. L. Powell (1973) Basalts from Iceland and along the Reykjanes ridge: Sr isotope geochemistry, *Nature (London), Phys. Sci., 246,* 104–107.

Haskin, L. A., M. A. Haskin, F. A. Frey, and T. R. Wildeman (1968) Relative and absolute terrestrial abundances of the rare earths. In L. H. Ahrens, ed., *Origin and distribution of the elements,* Pergamon Press, New York, p. 889–912.

Hazen, R. M. (1975) Effects of temperature and pressure on the crystal physics of olivine, Ph.D. thesis, Harvard University, Cambridge, Massachusetts, 264 p.

Heard, H. C. (1960) Transition from brittle fracture to ductile flow in Solenhofen limestone as a function of temperature, confining pressure, and interstitial fluid pressure. In D. Griggs and J. Handin, eds., Rock deformation, *Geol. Soc. Am. Mem. 79*, 193–244.

Heard, H. C., and W. W. Rubey (1966) Tectonic implications of gypsum dehydration, *Geol. Soc. Am. Bull., 77*, 741–760.

Hedge, C. E., and Z. E. Peterman (1970) The strontium isotopic composition of basalts from the Gordo and Juan de Fuca rises, northeastern Pacific Ocean, *Contrib. Mineral. Petrol., 27*, 114–120.

Heirtzler, J. R., G. O. Dickson, E. M. Herron, W. C. Pitman III, and X. Le Pichon (1968) Marine magnetic anomalies, geomagnetic field reversals, and motions of the ocean floor and continents, *J. Geophys. Res., 73*, 2119–2136.

Helmke, P. A., and L. A. Haskin (1973) Rare-earth elements, Co, Sc and Hf in the Steens Mountain basalts, *Geochim. Cosmochim. Acta, 37*, 1513–1529.

Herty, C. H., Jr., F. A. Hartgen, J. A. Heidish, K. Metcalfe, F. G. Norris, and M. B. Royer (1930) Temperature-viscosity relations in the lime–silica system, *Carnegie Inst. Technol., Coop. Bull. Min. Metall. Invest., 47*, 27 p.

Hess, H. H. (1960) Stillwater igneous complex, Montana, *Geol. Soc. Am. Mem. 80*, 230 p.

Hodges, F. N. (1974) The solubility of H_2O in silicate melts, *Carnegie Inst. Washington Yearb. 73*, 251–255.

Hofmann, A. (1974) Strontium diffusion in a basalt melt and implications for Sr isotope geochemistry and geochronology, *Carnegie Inst. Washington Yearb. 73*, 935–941.

Hofmann, A. W., and S. R. Hart (1975) An assessment of local and regional isotopic equilibrium in a partially molten mantle, *Carnegie Inst. Washington Yearb. 74*, 195–210.

Holmes, A. (1921) *Petrographic methods and calculations*, Thomas Murby & Co., London, 515 p.

Holmes, A. (1915) Radioactivity and the earth's thermal history, *Geol. Mag., 2*, 60–70, 102–111.

Holmes, A. (1927) Some problems of physical geology and the earth's thermal history, *Geol. Mag., 64*, 263–278.

Holmes, A. (1945) *Principles of physical geology*, The Ronald Press Company, New York, 532 p.

Holmes, A. (1965) *Principles of physical geology*, Th. Nelson and Sons, Ltd., London, 1288 p.

Hubbert, M. K., and W. W. Rubey (1959) Role of fluid pressure in mechanics of overthrust faulting, I, Mechanics of fluid-filled porous solids and its application to overthrust faulting, *Geol. Soc. Am. Bull., 70*, 115–166.

Hubbert, M. K., and D. G. Willis (1957) Mechanics of hydraulic fracturing, *Trans. AIME, 210*, 153–168.

Huckenholz, H. G. (1965) Der petrogenetische Werdegang der Klinopyroxene in den tertiären Vulkaniten der Hocheifel (Parts I and II), *Beitr. Mineral. Petrogr., 11*, 138–195, 415–448.

Huckenholz, H. G. (1966) Der petrogenetische Werdegang der Klinopyroxene in den tertiären Vulkaniten der Hocheifel (Part III), *Contrib. Mineral. Petrol., 12*, 73–95.

Huckenholz, H. G. (1973) The origin of fassaitic augite in the alkali basalt suite of the Hocheifel area, western Germany, *Contrib. Mineral. Petrol., 40*, 315–326.

Hulme, G. (1974) The interpretation of lava flow morphology, *Geophys. J. R. Astron. Soc., 39*, 361–383.

Hurley, P. M. (1968a) Absolute abundance and distribution of Rb, K and Sr in the earth, *Geochim. Cosmochim. Acta, 32*, 273–283.

Hurley, P. M. (1968b) Correction to: Absolute abundance and distribution of Rb, K and Sr in the earth, *Geochim. Cosmochim. Acta, 32,* 1025–1030.

Hytönen, K., and J. F. Schairer (1960) The system enstatite–anorthite–diopside, *Carnegie Inst. Washington Yearb. 59,* 71–72.

Hytönen, K., and J. F. Schairer (1961) The plane enstatite–anorthite–diopside and its relation to basalts, *Carnegie Inst. Washington Yearb. 60,* 125–141.

Iida, K. (1959) Earthquake energy and earthquake fault, *J. Earth Sci. Nagoya Univ., 7,* 98–107.

Irvine, T. N. (1970) Crystallization sequences in the Muskox intrusion and other layered intrusions, 1, Olivine–pyroxene–plagioclase relations, *Geol. Soc. S. Afr. Spec. Publ. 1,* 441–476.

Irving, A. J., and P. J. Wyllie (1975) Subsolidus and melting relationships for calcite, magnesite and the join $CaCO_3$–$MgCO_3$ to 36 kb, *Geochim. Cosmochim. Acta, 39,* 35–53.

Ito, K., and G. C. Kennedy (1967) Melting and phase relations in a natural peridotite to 40 kilobars, *Am. J. Sci., 265,* 519–538.

Ito, K., and G. C. Kennedy (1968) Melting and phase relations in the plane tholeiite–lherzolite–nepheline basanite to 40 kilobars with geological implications. *Contrib. Mineral. Petrol., 19,* 177–211.

Ito, K., and G. C. Kennedy (1974) The composition of liquids formed by partial melting of eclogites at high temperatures and pressures, *J. Geol., 82,* 383–392.

Jackson, E. D., E. A. Silver, and G. B. Dalrymple (1972) Hawaiian–Emperor chain and its relation to Cenozoic circumpacific tectonics, *Geol. Soc. Am. Bull., 83,* 601–617.

Jackson, E. D., and T. L. Wright (1970) Xenoliths in the Honolulu volcanic series, Hawaii, *J. Petrol., 11,* 405–430.

Jaggar, T. A. (1917) Volcanic investigations at Kilauea, *Am. J. Sci., 44,* 165–220.

Johannsen, A. (1939) *A descriptive petrography of the igneous rocks,* Vol. 1, University of Chicago Press, Chicago, 318 p.

Joly, J. (1909) *Radioactivity and geology,* Archibald Constable and Company, Ltd., London, 287 p.

Joly, J. (1930) *The surface-history of the earth,* Clarendon Press, Oxford, 211 p.

Jordan, T. H. (1975) The continental tectosphere, *Rev. Geophys. Space Phys., 13,* No. 3, 1–12.

Joule, J. P., and J. Thomson (1854) On the thermal effects of fluids in motion—No. II, *Proc. R. Soc. London,* No. 7, 127–130.

Judd, J. W. (1886) On the gabbros, dolerites and basalts of Tertiary age in Scotland and Ireland, *Q. J. Geol. Soc. London, 42,* 49–97.

Kaula, W. M. (1963) Elastic models of the mantle corresponding to variations in the external gravity field, *J. Geophys. Res., 68,* 4967–4978.

Kawada, K. (1966) Studies of the thermal state of the earth, The 17th paper: Variation of thermal conductivity of rocks, Pt. 2, *Bull. Earthquake Res. Inst. Tokyo Univ., 44,* 1071–1091.

Kennedy, G. C. (1959) The origin of continents, mountain ranges, and ocean basins, *Am. Sci., 47,* 491–504.

Kirby, S. H., and C. B. Raleigh (1973) Mechanisms of high-temperature, solid-state flow in minerals and ceramics and their bearing on the creep behavior of the mantle, *Tectonophysics, 19,* 165–194.

Knopoff, L. (1972) Observation and inversion of surface-wave dispersion, *Tectonophysics, 13,* 497–519.

Koyanagi, R. Y., and E. T. Endo (1971) Hawaiian seismic events during 1969, *U.S. Geol. Surv. Prof. Pap. 750-C*, C158–C164.

Kozu, S., and J. Ueda (1933) Thermal expansion of diopside, *Proc. Imp. Acad. (Tokyo), 9*, 317–319.

Kretz, R. (1961) Some applications of thermodynamics to coexisting minerals of variable compositions. Examples: orthopyroxene–clinopyroxene and orthopyroxene–garnet, *J. Geol., 69*, 361–387.

Kubota, S., and E. Berg (1967) Evidence for magma in the Katmai volcanic range, *Bull. Volcanol., 31*, 175–214.

Kuno, H. (1959) Origin of Cenozoic petrographic provinces of Japan and surrounding areas, *Bull. Volcanol., 20*, 37–76.

Kuno, H. (1967) Volcanological and petrological evidences regarding the nature of the upper mantle. In T. F. Gaskell, ed., *The earth's mantle*, Academic Press, New York, p. 89–110.

Kuno, H. (1969) Plateau basalts. In P. J. Hart, ed., The earth's crust and upper mantle, *Geophys. Monogr., Am. Geophys. Union, 13*, 495–501.

Kuno, H., and K.-I. Aoki (1970) Chemistry of ultramafic nodules and their bearing on the origin of basaltic magmas, *Phys. Earth Planet. Inter., 3*, 273–301.

Kushiro, I. (1964) Coexistence of nepheline and enstatite at high pressures, *Carnegie Inst. Washington Yearb. 64*, 109–112.

Kushiro, I. (1968) Compositions of magmas formed by partial zone melting of the earth's upper mantle, *J. Geophys. Res., 73*, 619–634.

Kushiro, I. (1969a) Clinopyroxene solid solutions formed by reactions between diopside and plagioclase at high pressures, *Mineral. Soc. Am. Spec. Pap. 2*, 179–191.

Kushiro, I. (1969b) The system forsterite–diopside–silica with and without water at high pressures, *Am. J. Sci., Schairer Vol., 267A*, 269–294.

Kushiro, I. (1970) Stability of amphibole and phlogopite in the upper mantle, *Carnegie Inst. Washington Yearb. 68*, 245–247.

Kushiro, I. (1972a) Determination of liquidus relations in synthetic silicate systems with electron probe analysis: the system forsterite–diopside–silica at 1 atmosphere, *Am. Mineral., 57*, 1260–1271.

Kushiro, I. (1972b) Effect of water on the compositions of magmas formed at high pressures, *J. Petrol., 13*, 311–334.

Kushiro, I. (1973a) Origin of some magmas in oceanic and circum-oceanic regions, *Tectonophysics, 17*, 211–222.

Kushiro, I. (1973b) Partial melting of garnet lherzolites from kimberlite at high pressures. In P. H. Nixon, ed., *Lesotho kimberlites*, Lesotho National Development Corporation, Maseru, Lesotho, p. 294–299.

Kushiro, I. (1974) The system forsterite–anorthite–albite–silica–H_2O at 15 kbar and the genesis of andesitic magmas in the upper mantle, *Carnegie Inst. Washington Yearb. 73*, 244–248.

Kushiro, I., Y. Syono, and S. Akimoto (1968a) Melting of a peridotite at high pressures and high water pressures, *J. Geophys. Res., 73*, 6023–6029.

Kushiro, I., and R. N. Thompson (1972) Origin of some abyssal tholeiites from the mid-Atlantic ridge, *Carnegie Inst. Washington Yearb. 71*, 403–406.

Kushiro, I., and H. S. Yoder, Jr. (1966) Anorthite–forsterite and anorthite–enstatite reactions and their bearing on the basalt–eclogite transformation, *J. Petrol., 7*, 337–362.

Kushiro, I., and H. S. Yoder, Jr. (1969) Melting of forsterite and enstatite at high pressures under hydrous conditions, *Carnegie Inst. Washington Yearb. 67*, 153–158.

Kushiro, I., and H. S. Yoder, Jr. (1974) Formation of eclogite from garnet lherzolite liquidus relations in a portion of the system $MgSiO_3-CaSiO_3-Al_2O_3$ at high pressures, *Carnegie Inst. Washington Yearb. 73*, 266–269.

Kushiro, I., H. S. Yoder, Jr., and B. O. Mysen (In press) Viscosity of basalt and andesite melts at high pressures.

Kushiro, I., H. S. Yoder, Jr., and M. Nishikawa (1968b) Effect of water on the melting of enstatite, *Geol. Soc. Am. Bull., 79*, 1685–1692.

Lacroix, A. (1917) Sur la transformation de quelques roches éruptives basiques en amphibolites, *C. R. Acad. Sci., 164*, 969–974.

Lambert, I. B., and P. J. Wyllie (1970) Low-velocity zone of the earth's mantle: incipient melting caused by water, *Science, 169*, 764–766.

Larsen, E. S. (1909) Relation between the refractive index and the density of some crystallized silicates, *Am. J. Sci., 28*, 263–274.

Larsen, E. S., and H. Berman (1934) The microscopic determination of the nonopaque minerals, *U.S. Geol. Surv. Bull. 848*, 266 p.

Lee, W. H. K. (1967) Thermal history of the earth, Ph.D. thesis in planetary and space physics, University of California at Los Angeles. 344 p.

Lindemann, F. A. (1910) Über die Berechnung Molekularer Eigenfrequenzen, *Phys. Z., 11*, 609–612.

Lindsley, D. H. (1967) Pressure–temperature relations in the system $FeO-SiO_2$, *Carnegie Inst. Washington Yearb. 65*, 226–230.

Liu, L. (1974) Silicate perovskite from phase transformations of pyrope–garnet at high pressure and temperature, *Geophys. Res. Lett., 1*, 277–280.

Lliboutry, L. (1971) Permeability, brine content and temperature of temperate ice, *J. Geol., 10*, 15–29.

Lovering, J. F., and J. W. Morgan (1964) Uranium and thorium abundances in stony meteorites, *J. Geophys. Res., 69*, 1979–1988.

Lubimova, E. A. (1958) Thermal history of the earth with consideration of the variable thermal conductivity of its mantle, *Geophys. J. R. Astron. Soc., 1*, 115–134.

Lubimova, E. A. (1969) Thermal history of the earth. In P. J. Hart, ed., The earth's crust and upper mantle, *Geophys. Monogr., Am. Geophys. Union, 13*, 63–77.

Lupton, J. E., and H. Craig (1975) Excess ^3He in oceanic basalts: evidence for terrestrial primordial helium, *Earth Planet. Sci. Lett., 26*, 133–139.

Lure, M. L., and V. L. Masaitis (1964) Principal features of the geology and petrology of the trap formations of the Siberian Platform, *Int. Geol. Congr., 22nd Session, Plateau Basalts*, Nauka, Moscow.

Luth, W. C. (1967) Studies in the system $KAlSiO_4-Mg_2SiO_4-SiO_2-H_2O$: I, Inferred phase relations and petrologic applications, *J. Petrol., 8*, 372–416.

Maaløe, S. (1973) The significance of the melting interval of basaltic magmas at various pressures, *Geol. Mag., 110*, 103–112.

Macdonald, G. A. (1972) *Volcanoes*, Prentice-Hall, Inc., Englewood Cliffs, New Jersey, 510 p.

MacDonald, G. J. F. (1959) Calculations on the thermal history of the earth, *J. Geophys. Res., 64*, 1967–2000.

MacGregor, I. D. (1968) Mafic and ultramafic inclusions as indicators of the depth of origin of basaltic magmas, *J. Geophys. Res., 73*, 3737–3745.

MacGregor, I. D. (1974) The system $MgO-Al_2O_3-SiO_2$: solubility of Al_2O_3 in enstatite from spinel and garnet peridotite compositions, *Am. Mineral., 59*, 110–119.

Machado, F. (1974) The search for magmatic reservoirs. In L. Civetta, P. Gasparini, G. Luongo, and A. Rapolla, eds., *Physical volcanology*, American Elsevier Scientific Publishing Company, Inc., New York, p. 255–273.

Manson, V. (1967) Geochemistry of basaltic rocks: major elements. In H. H. Hess and A. Poldervaart, eds., *Basalts: the Poldervaart treatise on rocks of basaltic composition*, Vol. 1, Interscience Publishers, New York, p. 215–269.

Mao, H. K., and P. M. Bell (1971) Generation of magma along faults and stress dislocations in the earth, *Carnegie Inst. Washington Yearb. 70*, 229–233.

Mao, H. K., and P. M. Bell (1976) High-pressure physics: the 1-megabar mark on the ruby R_1 static pressure scale, *Science, 191*, 851–852.

Mao, H. K., and J. F. Schairer (1970) Quenching experiments in the system jadeite ($NaAlSi_2O_6$)–forsterite (Mg_2SiO_4) and jadeite ($NaAlSi_2O_6$)–anorthite ($CaAl_2Si_2O_8$), *Carnegie Inst. Washington Yearb. 68*, 221–222.

Marsh, B. D., and I. S. E. Carmichael (1974) Benioff zone magmatism, *J. Geophys. Res., 79*, 1196–1206.

Masuda, A., and I. Kushiro (1970) Experimental determination of partition coefficients of ten rare earth elements and barium between clinopyroxene and liquid in synthetic silicate systems at 20 kilobar pressure, *Contrib. Mineral. Petrol., 26*, 42–49.

McBirney, A. R. (1963) Conductivity variations and terrestrial heat-flow distribution, *J. Geophys. Res., 68*, 6323–6329.

McBirney, A. R. (1967) Genetic relations of volcanic rocks of the Pacific Ocean, *Geol. Rundsch., 57*, 21–33.

McBirney, A. R., J. F. Sutter, H. R. Naslund, K. G. Sutton, and C. M. White (1974) Episodic volcanism in the Central Oregon Cascade Range, *Geology, 2*, 585–589.

McGetchin, T. R. (1975) Solid earth geosciences research activities at LASL, July 1–December 31, 1974, *Prog. Rep. LA-5956-PR*, Los Alamos Scientific Laboratories, U.S. Energy Research and Development Administration, 93 p.

McKenzie, D. P., and J. G. Sclater (1968) Heat flow inside the island arcs of the northwestern Pacific, *J. Geophys. Res., 73*, 3173–3179.

McLean, D. (1957) *Grain boundaries in metals*, Oxford University Press, Clarendon, 346 p.

McQuarrie, M. (1954) Thermal conductivity: VII, Analysis of variation of conductivity with temperature for Al_2O_3, BeO, and MgO, *J. Am. Ceram. Soc., 37*, 91–95.

Means, W. D., and P. F. Williams (1974) Compositional differentiation in an experimentally deformed salt–mica specimen, *Geology, 2*, 15–16.

Mehnert, K. R., W. Busch, and G. Schneider (1973) Initial melting at grain boundaries of quartz and feldspar in gneisses and granulites, *Neues Jahrb. Mineral. Monatsh.*, 165–183.

Menard, H. W. (1964) *Marine geology of the Pacific*, McGraw-Hill Book Company, New York, 271 p.

Menard, H. W. (1969) The deep-ocean floor, *Sci. Am., 221*, 127–142.

Menzies, M. (1973) Mineralogy and partial melt textures within an ultramafic–mafic body, Greece, *Contrib. Mineral. Petrol., 42*, 273–285.

Mercier, J.-C., and N. L. Carter (1975) Pyroxene geotherms, *J. Geophys. Res., 80*, 3349–3362.

Merrill, R. B., and P. J. Wyllie (1975) Kaersutite and kaersutite eclogite from Kakanui, New Zealand—water-excess and water-deficient melting to 30 kilobars, *Geol. Soc. Am. Bull., 86*, 555–570.

Millhollen, G. L., A. J. Irving, and P. J. Wyllie (1974) Melting interval of peridotite with 5.7 per cent water to 30 kilobars, *J. Geol., 82*, 575–587.

Modreski, P. J., and A. L. Boettcher (1972) The stability of phlogopite + enstatite at high pressures: a model for micas in the interior of the earth, *Am. J. Sci., 272*, 852–869.

Mogi, K. (1958) Relations between the eruptions of various volcanoes and the deformations of the ground surfaces around them, *Bull. Earthquake Res. Inst. Tokyo Univ., 36,* 99–134.

Mogi, K. (1965) Deformation and fracture of rocks under confining pressure, (2) Elasticity and plasticity of some rocks, *Bull. Earthquake Res. Inst. Tokyo Univ., 43,* 349–379.

Mogi, K. (1966) Pressure dependence of rock strength and transition from brittle fracture to ductile flow, *Bull. Earthquake Res. Inst. Tokyo Univ., 44,* 215–232.

Moorbath, S., R. K. O'Nions, R. J. Pankhurst, N. H. Gale, and V. R. McGregor (1972) Further rubidium–strontium age determinations on the very early Precambrian rocks of the Godthaab District, West Greenland, *Nature (London), Phys. Sci., 240,* 78–82.

Moore, P. B., and J. V. Smith (1970) Crystal structure of β-Mg_2SiO_4: crystal–chemical and geophysical implications, *Phys. Earth Planet. Inter., 3,* 166–177.

Morey, G. W. (1922) The development of pressure in magmas as a result of crystallization, *J. Wash. Acad. Sci., 12,* 219–230.

Morgan, W. J. (1971) Convection plumes in the lower mantle, *Nature (London), 230,* 42–43.

Morgan, W. J. (1972) Deep mantle convection plumes and plate motions, *Bull. Am. Assoc. Pet. Geol., 56,* 203–213.

Morimoto, N., S.-I. Akimoto, K. Koto, and M. Tokonami (1970) Crystal structures of high pressure modifications of Mn_2GeO_4 and Co_2SiO_4, *Phys. Earth Planet. Inter., 3,* 161–165.

Morse, S. A. (1970) Alkali feldspars with water at 5 kb pressure, *J. Petrol., 11,* 221–251.

Murase, T., and A. R. McBirney (1970) Thermal conductivity of lunar and terrestrial igneous rocks in their melting range, *Science, 170,* 165–167.

Murase, T., and A. R. McBirney (1973) Properties of some common igneous rocks and their melts at high temperatures, *Geol. Soc. Am. Bull., 84,* 3563–3592.

Murase, T., and T. Suzuki (1966) Ultrasonic velocity of longitudinal waves in molten rocks, *J. Fac. Sci. Hokkaido Univ., Ser. 7, 2,* 273–285.

Mysen, B. O. (1973) Melting in a hydrous mantle: Phase relations of mantle peridotite with controlled water and oxygen fugacities, *Carnegie Inst. Washington Yearb. 72,* 467–478.

Mysen, B. O. (1975) Solubility of volatiles in silicate melts at high pressure and temperature: the role of carbon dioxide and water in feldspar, pyroxene, and feldspathoid melts, *Carnegie Inst. Washington Yearb. 74,* 454–468.

Mysen, B. O. (1976a) Experimental determination of some geochemical parameters relating to conditions of equilibration of peridotite in the upper mantle, *Am. Mineral., 61,* 677–683.

Mysen, B. O. (1976b) Partitioning of samarium and nickel between olivine, orthopyroxene, and liquid: preliminary data at 20 kbar and 1025°C, *Earth Planet. Sci. Lett., 31,* 1–7.

Mysen, B. O., and A. L. Boettcher (1975) Melting of a hydrous mantle: I, Phase relations of natural peridotite at high pressures and temperatures with controlled activities of water, carbon dioxide and hydrogen, *J. Petrol., 16,* 520–548.

Nakamura, Y., and I. Kushiro (1974) Composition of the gas phase in Mg_2SiO_4–SiO_2–H_2O at 15 kbar, *Carnegie Inst. Washington Yearb. 73,* 255–258.

Nesbitt, R. W. (1971) Skeletal crystal forms in the ultramafic rocks of the Yilgarn block, western Australia; evidence for an Archean ultramafic liquid, *Spec. Publ. Geol. Soc. Aust., 3,* 331–347.

Nesterenko, G. V., and A. I. Almukhamedov (1973) *The geochemistry of differentiated traps (Siberian Platform)*, Nauka, Moscow, 198 p.

Nettleton, L. L. (1936) Fluid mechanics of salt domes. In D. C. Barton and G. Sawtelle, eds., *Gulf Coast Oil Fields; a Symposium on the Gulf Coast Cenozoic*, American Association of Petroleum Geologists, Tulsa, Oklahoma, p. 79–108.

Nier, A. O. (1950) A redetermination of the relative abundances of the isotopes of carbon, nitrogen, oxygen, argon, and potassium, *Phys. Rev., 77*, 789–793.

Nockolds, S. R. (1954) Average chemical compositions of some igneous rocks, *Geol. Soc. Am. Bull., 65*, 1007–1032.

Nur, A. (1972) Dilatancy, pore fluids, and premonitory variations of t_s/t_p travel times, *Bull. Seismol. Soc. Am., 62*, 1217–1222.

Nur, A. (1974) Tectonophysics: the study of relations between deformation and forces in the earth. In *Advances in rock mechanics*, Vol. 1, Part A, National Academy of Sciences, Washington, D.C., p. 243–317.

Nutting, P. G. (1929) Deformation and temperature, *J. Wash. Acad. Sci., 19*, 109–115.

Nye, J. F. (1967) Theory of regelation, *Philos. Mag., 16*, 1249–1266.

Nye, J. F., and F. C. Frank (1973) Hydrology of the intergranular veins in a temperate glacier. In Symposium on the hydrology of glaciers, *Assoc. Int. Hydrologie Scientifique Publ. 95*, 157–161.

Obata, M. (1975) The solubility of Al_2O_3 in orthopyroxenes in spinel and plagioclase peridotites and spinel pyroxenite. In *Extended abstracts: International Conference on Geothermometry and Geobarometry*, Pennsylvania State University, University Park, Pennsylvania, p. 1–2.

Obata, M., S. Banno, and T. Mori (1974) The iron–magnesium partitioning between naturally occurring coexisting olivine and Ca-rich clinopyroxene: an application of the simple mixture model to olivine solid solution, *Bull. Soc. Fr. Mineral. Cristallogr., 97*, 101–107.

O'Hara, M. J. (1961) Zoned ultrabasic and basic gneiss masses in the early Lewisian metamorphic complex at Scourie, Sutherland, *J. Petrol., 2*, 248–276.

O'Hara, M. J. (1965) Primary magmas and the origin of basalts, *Scott. J. Geol., 1*, 19–40.

O'Hara, M. J. (1968a) Are ocean floor basalts primary magma? *Nature (London), 220*, 683–686.

O'Hara, M. J. (1968b) The bearing of phase equilibria studies in synthetic and natural systems on the origin and evolution of basic and ultrabasic rocks, *Earth-Sci. Rev., 4*, 69–133.

O'Hara, M. J., M. J. Saunders, and E. L. P. Mercy (1975) Garnet–peridotite, primary ultrabasic magma and eclogite; interpretation of upper mantle processes in kimberlite, *Phys. Chem. Earth, 9*, 571–604.

O'Hara, M. J., and H. S. Yoder, Jr. (1967) Formation and fractionation of basic magmas at high pressures, *Scott. J. Geol., 3*, 67–117.

O'Nions, R. K., and R. J. Pankhurst (1974) Petrogenetic significance of isotope and trace element variations in volcanic rocks from the mid-Atlantic, *J. Petrol., 15*, 603–634.

Onuma, K., and K. Yagi (1967) The system diopside–akermanite–nepheline, *Am. Mineral., 52*, 227–243.

Orowan, E. (1960) Mechanism of seismic faulting. In D. Griggs and J. Handin, eds., Rock deformation, *Geol. Soc. Am. Mem. 79*, 323–345.

Osborn, E. F. (1942) The system $CaSiO_3$–diopside–anorthite, *Am. J. Sci., 240*, 751–788.

Osborn, E. F., and D. B. Tait (1952) The system diopside–forsterite–anorthite, *Am. J. Sci., Bowen Vol.*, 413–433.

Oxburgh, E. R. (1964) Petrological evidence for the presence of amphibole in the upper mantle and its petrogenetic and geophysical implications, *Geol. Mag., 101,* 1–19.

Oxburgh, E. R., and D. L. Turcotte (1970) Thermal structure of island arcs, *Geol. Soc. Am. Bull., 81,* 1665–1688.

Paterson, M. S. (1973) Nonhydrostatic thermodynamics and its geologic applications, *Rev. Geophys. Space Phys., 11,* 355–389.

Pfann, W. G. (1952) Principles of zone melting, *J. Met., 4,* 747–753.

Pfann, W. G. (1966) *Zone melting,* 2d Ed., John Wiley & Sons, Inc., New York, 310 p.

Phillips, W. J. (1972) Hydraulic fracturing and mineralization, *J. Geol. Soc. London, 218,* 337–359.

Poldervaart, A. (1962) Aspects of basalt petrology, *J. Geol. Soc. India, 3,* 1–14.

Porath, H., D. W. Oldenburg, and D. I. Gough (1970) Separation of magnetic variation fields and conductive structures in the western United States, *Geophys. J. R. Astron. Soc., 19,* 237–260.

Powell, M., and R. Powell (1974) An olivine–clinopyroxene geothermometer, *Contrib. Mineral. Petrol., 48,* 249–263.

Powell, R. W. (1954) Thermal conductivities of solid materials at high temperatures, *Research (London), 7,* 492–501.

Presnall, D. C. (1969) The geometrical analysis of partial fusion, *Am. J. Sci., 267,* 1178–1194.

Presnall, D. C., T. H. O'Donnell, and N. L. Brenner (1973) Cusps on solidus curves as controls for primary magma compositions: a mechanism for producing oceanic olivine tholeiites of uniform composition, *Abstr. with Programs (Geol. Soc. Am.), 5,* 771–772.

Press, F. (1965) Dimensions of the source region for small shallow earthquakes (unpublished VESIAC report).

Press, F. (1968) Earth models obtained by Monte Carlo inversion, *J. Geophys. Res., 73,* 5223–5234.

Press, F. (1969) The suboceanic mantle, *Science, 165,* 174–176.

Press, F. (1970) Earth models consistent with geophysical data, *Phys. Earth Planet. Inter., 3,* 3–22.

Pyke, D. R., A. J. Naldrett, and O. R. Eckstrand (1973) Archean ultramafic flows in Munro Township, Ontario, *Geol. Soc. Am. Bull., 84,* 955–978.

Råheim, A., and D. H. Green (1974) Experimental determination of the temperature and pressure dependence of the Fe–Mg partition coefficient for coexisting garnet and clinopyroxene, *Contrib. Mineral. Petrol., 48,* 179–203.

Raleigh, C. B., and M. S. Paterson (1965) Experimental deformation of serpentinite and its tectonic implications, *J. Geophys. Res., 70,* 3965–3985.

Ramberg, H. (1971) Temperature changes associated with adiabatic decompression in geological processes, *Nature (London), 234,* 539–540.

Ramberg, H. (1972) Mantle diapirism and its tectonic and magmagenetic consequences, *Phys. Earth Planet. Inter., 5,* 45–60.

Rankin, G. A., and F. E. Wright (1915) The ternary system $CaO–Al_2O_3–SiO_2$, *Am. J. Sci., 39,* 1–79.

Reay, A., and P. G. Harris (1964) The partial fusion of peridotite, *Bull. Volcanol., 27,* 115–127.

Reed, G. W., K. Kigoshi, and A. Turkevich (1960) Determinations of concentrations of heavy elements in meteorites by activation analysis, *Geochim. Cosmochim. Acta, 20,* 122–140.

Reitan, P. H. (1968) Frictional heat during metamorphism, *Lithos, 1*, 151–163, 268–274.

Richter, C. F. (1935) An instrumental earthquake magnitude scale, *Bull. Seismol. Soc. Am., 25*, 1–32.

Riecke, E. (1895) Über das Gleichgewicht zwischen einem festen, homogen deformirten Korper und einer flussigen Phase, insbesondere über die Depression des Schmelzpunktes durch einseitige Spannung, *Ann. Phys. (Leipzig), 54*, 731–738.

Riecke, E. (1912) Zur Erniedrigung des Schmelzpunktes durch einseitigen Zug oder Druck, *Zentralbl. Mineral. Geol. Palaeontol., 3*, 97–104.

Ringwood, A. E. (1958) Constitution of the mantle, 3, *Geochim. Cosmochim. Acta, 15*, 195–212.

Ringwood, A. E. (1962) A model for the upper mantle, *J. Geophys. Res., 67*, 857–867.

Ringwood, A. E. (1966) Mineralogy of the mantle. In P. M. Hurley, ed., *Advances in earth science*, MIT Press, Cambridge, p. 357–399.

Ringwood, A. E. (1969) Composition of the crust and upper mantle. In P. J. Hart, ed., The earth's crust and upper mantle, *Geophys. Monogr., Am. Geophys. Union, 13*, 1–17.

Ringwood, A. E. (1970) Phase transformations and the constitution of the mantle, *Phys. Earth Planet. Inter., 3*, 109–155.

Rittmann, A. (1962) *Volcanoes and their activity*, John Wiley & Sons, Inc., New York, 305 p.

Roberts, J. L. (1970) The intrusion of magma into brittle rocks. In G. Newall and N. Rast, eds., *Mechanism of igneous intrusion*, Gallery Press, Liverpool, p. 287–338.

Robie, R. A., and D. R. Waldbaum (1968) Thermodynamic properties of minerals and related substances at 298.15°K (25.0°C) and one atmosphere (1.013 bars) and at higher temperatures, *U.S. Geol. Surv. Bull. 1259*, 256 p.

Roeder, P. L. (1974) Paths of crystallization and fusion in systems showing ternary solid solution, *Am. J. Sci., 274*, 48–60.

Rogers, J. J. W., and J. A. S. Adams (1966) *Fundamentals of geology*, Harper and Row, New York, 424 p.

Ronov, A. B., and A. A. Yaroshevsky (1969) Chemical composition of the earth's crust. In P. J. Hart, ed., The earth's crust and upper mantle, *Geophys. Monogr., Am. Geophys. Union, 13*, 37–57.

Roscoe, R. (1952) The viscosity of suspensions of rigid spheres, *Br. J. Appl. Phys., 3*, 267–269.

Rosenhauer, M., and D. H. Eggler (1975) Solution of H_2O and CO_2 in diopside melt, *Carnegie Inst. Washington Yearb. 74*, 474–479.

Rubey, W. W. (1951) Geologic history of sea water, *Geol. Soc. Am. Bull., 62*, 1111–1147.

Rubey, W. W., and M. K. Hubbert (1959) Role of fluid pressure in mechanics of overthrust faulting, II, Overthrust belt in geosynclinal area of western Wyoming in light of fluid-pressure hypothesis, *Geol. Soc. Am. Bull., 70*, 167–205.

Runcorn, S. K. (1967) *International dictionary of geophysics*, Pergamon Press, Oxford, 1728 p.

Rutter, E. H. (1972) The influence of interstitial water on the rheological behaviour of calcite rocks, *Tectonophysics, 14*, 13–33.

Sacks, I. S., and H. Okada (1974) A comparison of the anelasticity structure beneath western South America and Japan, *Phys. Earth Planet. Inter., 9*, 211–219.

Sanford, A. R., O. Alptekin, and T. R. Topposada (1973) Use of reflection phases on microearthquake seismograms to map an unusual discontinuity beneath the Rio Grande rift, *Bull. Seismol. Soc. Am., 63*, 2021–2034.

Savage, J. C. (1969) The mechanics of deep-focus faulting, *Tectonophysics, 8*, 115–127.

230 References

Savanick, G. A., and D. I. Johnson (1972) Adhesion at crystalline interfaces in rock, *U.S. Bur. Mines Rep. Invest. 7709*, 16 p.

Scarfe, C. M. (1973) Viscosity of basic magmas at varying pressure, *Nature (London), Phys. Sci., 241*, 101–102.

Scarfe, C. M., D. K. Paul, and P. G. Harris (1972) Melting experiments at one atmosphere on two ultramafic nodules, *Neues Jahrb. Mineral. Monatsh.*, 469–476.

Schairer, J. F. (1942) The system $CaO-FeO-Al_2O_3-SiO_2$: I, Results of quenching experiments on five joins, *J. Am. Ceram. Soc., 25*, 241–274.

Schairer, J. F. (1954) The system $K_2O-MgO-Al_2O_3-SiO_2$: I, Results of quenching experiments on four joins in the tetrahedron cordierite–forsterite–leucite–silica and on the join cordierite–mullite–potash feldspar, *J. Am. Ceram. Soc., 37*, 501–533.

Schairer, J. F. (1957) Melting relations of the common rock-forming oxides, *J. Am. Ceram. Soc., 40*, 215–235.

Schairer, J. F., and N. L. Bowen (1942) The binary system $CaSiO_3$–diopside and the relations between $CaSiO_3$ and akermanite, *Am. J. Sci., 240*, 725–742.

Schairer, J. F., and N. Morimoto (1958) Systems with rock-forming olivines, pyroxenes, and feldspars, *Carnegie Inst. Washington Yearb. 57*, 212–213.

Schairer, J. F., and N. Morimoto (1959) The system forsterite–diopside–silica–albite, *Carnegie Inst. Washington Yearb. 58*, 113–118.

Schairer, J. F., C. E. Tilley, and G. M. Brown (1968) The join nepheline–diopside–anorthite and its relation to alkali basalt fractionation, *Carnegie Inst. Washington Yearb. 66*, 467–471.

Schairer, J. F., K. Yagi, and H. S. Yoder, Jr. (1962) The system nepheline–diopside, *Carnegie Inst. Washington Yearb. 61*, 96–98.

Schairer, J. F., and H. S. Yoder, Jr. (1960a) The nature of residual liquids from crystallization, with data on the system nepheline–diopside–silica, *Am. J. Sci., Bradley Vol., 258A*, 273–283.

Schairer, J. F., and H. S. Yoder, Jr. (1960b) The system forsterite–nepheline–diopside, *Carnegie Inst. Washington Yearb. 59*, 70–71.

Schairer, J. F., and H. S. Yoder, Jr. (1961) Crystallization in the system nepheline–forsterite–silica at one atmosphere pressure, *Carnegie Inst. Washington Yearb. 60*, 141–144.

Schairer, J. F., and H. S. Yoder, Jr. (1964) Crystal and liquid trends in simplified alkali basalts, *Carnegie Inst. Washington Yearb. 63*, 65–74.

Schairer, J. F., and H. S. Yoder, Jr. (1967) The system albite–anorthite–forsterite at 1 atmosphere, *Carnegie Inst. Washington Yearb. 65*, 204–209.

Schairer, J. F., and H. S. Yoder, Jr. (1969) The join albite–anorthite–akermanite, *Carnegie Inst. Washington Yearb. 67*, 104–105.

Schairer, J. F., and H. S. Yoder, Jr. (1970) Critical planes and flow sheet for a portion of the system $CaO-MgO-Al_2O_3-SiO_2$ having petrological applications, *Carnegie Inst. Washington Yearb. 68*, 202–214.

Schairer, J. F., H. S. Yoder, Jr., and C. E. Tilley (1965) Behavior of melilites in the join gehlenite–soda melilite–akermanite at one-atmosphere pressure, *Carnegie Inst. Washington Yearb. 64*, 95–100.

Schilling, J.-G. (1966) *Rare earth fractionation in Hawaiian volcanic rocks*, Massachusetts Institute of Technology thesis, 390 p.

Schilling, J.-G. (1971) Sea-floor evolution: rare earth evidence, *Philos, Trans. R. Soc. London, Ser. A, 268*, 663–706.

Schilling, J.-G. (1975) Rare-earth variations across "normal segments" of the Reykjanes Ridge, 60°–53°N, Mid-Atlantic Ridge, 29°S, and East Pacific Rise, 2°–19°S, and evidence on the composition of the underlying low-velocity layer, *J. Geophys. Res., 80,* 1459–1473.

Schilling, J.-G., and E. Bonatti (1975) East Pacific ridge (2°S–19°S) versus Nazca intraplate volcanism: rare-earth evidence, *Earth Planet. Sci. Lett., 25,* 93–102.

Schilling, J.-G., and J. W. Winchester (1967) Rare-earth fractionation and magmatic processes. In S. K. Runcorn, ed., *Mantles of the earth and terrestrial planets,* Interscience Publishers, New York, p. 267–283.

Schmitt, R. A., R. H. Smith, J. E. Lasch, A. W. Mosen, D. A. Olehy, and J. Vasilevskis (1963) Abundances of the fourteen rare-earth elements, scandium, and yttrium in meteoritic and terrestrial matter, *Geochim. Cosmochim. Acta, 27,* 577–622.

Schmitt, R. A., R. H. Smith, and D. A. Olehy (1964) Rare-earth, yttrium and scandium abundances in meteoritic and terrestrial matter—II, *Geochim. Cosmochim. Acta, 28,* 67–86.

Sclater, J. G., and J. Francheteau (1970) The implications of terrestrial heat flow observations on current tectonic and geochemical models of the crust and upper mantle of the earth, *Geophys. J. R. Astron. Soc., 20,* 509–542.

Seitz, M. G., and I. Kushiro (1974) Melting relations of the Allende meteorite, *Science, 183,* 954–957.

Shaw, H. R. (1965) Comments on viscosity, crystal settling, and convection in granitic magmas, *Am. J. Sci., 263,* 120–152.

Shaw, H. R. (1969) Rheology of basalt in the melting range, *J. Petrol., 10,* 510–535.

Shaw, H. R. (1970) Earth tides, global heat flow, and tectonics, *Science, 168,* 1084–1087.

Shaw, H. R. (1973) Mantle convection and volcanic periodicity in the Pacific; evidence for Hawaii, *Geol. Soc. Am. Bull., 84,* 1505–1526.

Shaw, H. R. (1974) Diffusion of H_2O in granitic liquids: Part I, Experimental data; Part II, Mass transfer in magma chambers. In A. W. Hofmann, B. J. Giletti, H. S. Yoder, Jr., and R. A. Yund, eds., Geochemical transport and kinetics, *Carnegie Inst. Washington Publ. 634,* 139–170.

Shaw, H. R., and E. D. Jackson (1973) Linear island chains in the Pacific: result of thermal plumes or gravitational anchors? *J. Geophys. Res., 78,* 8634–8652.

Shaw, H. R., and D. A. Swanson (1970) Eruption and flow rates of flood basalts. In E. H. Gilmour and D. Stradling, eds., *Proceedings of the Second Columbia River Basalt Symposium,* EWSC Press, Cheney, Washington, p. 271–299.

Shaw, H. R., T. L. Wright, D. L. Peck, and R. Okamura (1968) The viscosity of basaltic magma; an analysis of field measurements in Makaopuhi Lava Lake, Hawaii, *Am. J. Sci., 266,* 225–264.

Shelley, D. (1974) Mechanical production of metamorphic banding—a critical appraisal, *Geol. Mag., 111,* 287–292.

Shimazu, Y. (1959) A thermodynamical aspect of the earth's interior—physical interpretation of magmatic differentiation process, *J. Earth Sci., Nagoya Univ., 7,* 1–34.

Shimazu, Y. (1961) Physical theory of generation, upward transfer, differentiation, solidification, and explosion of magmas, *J. Earth Sci., Nagoya Univ., 9,* 185–223.

Shimizu, N., and I. Kushiro (1973) Trace element content of liquid formed by partial melting of a garnet lherzolite at high pressures: a preliminary report, *Carnegie Inst. Washington Yearb. 72,* 270–272.

Shimizu, N., and I. Kushiro (1975) The partitioning of rare earth elements between garnet and liquid at high pressures: preliminary experiments, *Geophys. Res. Lett., 2,* 413–416.

Shimozuru, D. (1963) Geophysical evidence for suggesting the existence of molten pockets in the earth's upper mantle, *Bull. Volcanol., 26,* 181–195.

Shreve, R. L. (1972) Movement of water in glaciers, *J. Geol., 11,* 205–214.

Skinner, B. J. (1956) Physical properties of end-members of the garnet group, *Am. Mineral., 41,* 428–436.

Skinner, B. J. (1962) Thermal expansion of ten minerals, *U.S. Geol. Surv. Prof. Pap. 450D,* 109–112.

Sleep, N. H. (1974) Segregation of magma from a mostly crystalline mush, *Geol. Soc. Am. Bull., 85,* 1225–1232.

Smith, C. S. (1948) Grains, phases, and interfaces: An interpretation of microstructure, *Trans.* AIME, *175,* 15–51.

Sobolev, N. V., Jr., I. K. Kuznetsova, and N. I. Zyuzin (1968) The petrology of grospydite xenoliths from the Zagadochnaya kimberlite pipe in Yakutia, *J. Petrol., 9,* 253–280.

Sorby, H. C. (1863) Über Kalkstein-Geschiebe mit Eindrucken, *Neues Jahrb. Mineral. Geol. Palaeontol.,* 801–807.

Sorby, H. C. (1879) The anniversary address of the President, *Proc. Geol. Soc. London, 35,* 56–95.

Spetzler, H., and D. L. Anderson (1968) The effect of temperature and partial melting on velocity and attenuation in a simple binary system, *J. Geophys. Res., 73,* 6051–6060.

Spry, Alan (1962) The origin of columnar jointing, particularly in basalt flows, *J. Geol. Soc. Aust., 8,* 191–216.

Starr, A. T. (1928) Slip in a crystal and rupture in a solid due to shear, *Proc. Cambridge Philos. Soc., 24,* 489–500.

Steeples, D. W., and H. M. Iyer (1976) Low-velocity zone under Long Valley as determined from teleseismic events, *J. Geophys. Res., 81,* 849–860.

Stocker, R. L., and M. F. Ashby (1973) On the rheology of the upper mantle, *Rev. Geophys. Space Phys., 2,* 391–426.

Subramanian, A. P., and Y. S. Sahasrabudhe (1964) Geology of greater Bombay and Aurangabad–Ellora–Ajanta area, *Int. Geol. Congr., 22nd Session, New Delhi, India, Guide to Excursions Nos. A-13 and C-10,* p. 1–12.

Sun, S.-S., M. Tatsumoto, and J.-G. Schilling (1975) Mantle plume mixing along the Reykjanes Ridge axis: lead isotopic evidence, *Science, 190,* 143–147.

Swanson, D. A. (1972) Magma supply rate at Kilauea Volcano, 1952–1971, *Science, 175,* 169–170.

Tarakanov, R. Z., and N. V. Leviy (1968) A model for the upper mantle with several channels of low velocity and strength. In L. Knopoff, C. L. Drake, and P. J. Hart, eds., The crust and upper mantle of the Pacific area, *Geophys. Monogr., Am. Geophys. Union, 12,* 43–50.

Tatsumoto, M. (1966) Genetic relations of oceanic basalts as indicated by lead isotopes, *Science, 153,* 1094–1101.

Tatsumoto, M., C. E. Hedge, and A. E. J. Engel (1965) Potassium, rubidium, strontium, thorium and uranium in oceanic tholeiitic basalt, *Science, 150,* 886–888.

Taylor, H. P., Jr. (1974) The application of oxygen and hydrogen isotope studies to problems of hydrothermal alteration and ore deposition, *Econ. Geol., 69,* 843–883.

Tera, F., D. A. Papanastassiou, and G. J. Wasserburg (1974) Isotopic evidence for a terminal lunar cataclysm, *Earth Planet. Sci. Lett., 22,* 1–21.

Thompson, R. N. (1972) Melting behavior of two Snake River lavas at pressures up to 35 kb, *Carnegie Inst. Washington Yearb. 71,* 406–410.

Thomson, J. (1862) On crystallization and liquefaction, as influenced by stresses tending to change of form in the crystals, *Proc. R. Soc. London, 11,* 473–481. [Reproduced in *Phil. Mag., 24,* 395–401 (1862).]

Thorarinsson, S. (1954) The tephra-fall from Hekla on March 29th 1947. In T. Einarsson, G. Kjartansson, and S. Thorarinsson, eds., *The eruption of Hekla 1947–1948,* Vol. II, No. 3, Societas Scientiarum Islandica, Reykjavik, 68 p.

Tikhonov, A. N., E. A. Lubimova, and V. K. Vlasov (1970) On the evolution of melting zones in the thermal history of the earth, *Phys. Earth Planet. Inter., 2,* 326–331.

Tilley, C. E., H. S. Yoder, Jr., and J. F. Schairer (1964) New relations on melting of basalts, *Carnegie Inst. Washington Yearb. 63,* 92–97.

Tocher, D. (1958) Earthquake energy and ground breakage, *Bull. Seismol. Soc. Am., 48,* 147–153.

Toksöz, M. N., J. W. Minear, and B. R. Julian (1971) Temperature field and geophysical effects of a downgoing slab, *J. Geophys. Res., 76,* 1113–1138.

Tolland, H. G., and R. G. J. Strens (1972) Electrical conduction in physical and chemical mixtures; application to planetary mantles, *Phys. Earth Planet. Inter., 5,* 380–386.

Tolstikhin, I. N. (1975) Helium isotopes in the earth's interior and in the atmosphere: a degassing model of the earth, *Earth Planet. Sci. Lett., 26,* 88–96.

Tuthill, R. L. (1969) Effect of varying f_{O_2} on the hydrothermal melting and phase relations of basalt (abstract), *Eos, 50,* 355.

Tuttle, O. F., and N. L. Bowen (1958) Origin of granite in the light of experimental studies in the system $NaAlSi_3O_8-KAlSi_3O_8-SiO_2-H_2O$, *Geol. Soc. Am. Mem. 74,* 153 p.

Ubbelohde, A. R. (1965) *Melting and crystal structure,* Clarendon Press, Oxford, 325 p.

Uffen, R. J. (1959) On the origin of rock magma, *J. Geophys. Res., 64,* 117–122.

Uffen, R. J., and A. M. Jessop (1963) The stress release hypothesis of magma formation, *Bull. Volcanol., 26,* 57–66.

Van Schmus, W. R., and J. A. Wood (1967) A chemical–petrologic classification for the chondritic meteorites, *Geochim. Cosmochim. Acta, 31,* 747–765.

Verhoogen, J. (1946) Volcanic heat, *Am. J. Sci., 244,* 745–771.

Verhoogen, J. (1954) Petrological evidence on temperature distribution in the mantle of the earth, *Trans. Am. Geophys. Union, 35,* 85–92.

Verhoogen, J. (1973) Possible temperatures in the oceanic upper mantle and the formation of magma, *Geol. Soc. Am. Bull., 84,* 515–522.

Vernon, R. H. (1968) Microstructures of high-grade metamorphic rocks at Broken Hill, Australia, *J. Petrol., 9,* 1–22.

Vernon, R. H. (1970) Comparative grain-boundary studies of some basic and ultrabasic granulites, nodules and cumulates, *Scott. J. Geol., 6,* 337–351.

Viljoen, M. J., and R. P. Viljoen (1969a) The geology and geochemistry of the lower ultramafic unit of the Onverwacht Group and a proposed new class of igneous rock. In The Upper Mantle Project, *Geol. Soc. S. Afr. Spec. Publ. 2,* 55–85.

Viljoen, M. J., and R. P. Viljoen (1969b) Evidence for the existence of a mobile extrusive peridotitic magma from the Komati Formation of the Onverwacht Group. In The Upper Mantle Project, *Geol. Soc. S. Afr. Spec. Publ. 2,* 87–112.

Viljoen, M. J., and R. P. Viljoen (1969c) Evidence for the composition of the primitive mantle and its products of partial melting from a study of the rocks of the Barberton Mountain land. In The Upper Mantle Project, *Geol. Soc. S. Afr. Spec. Publ. 2,* 275–295.

Vinogradov, A. P., A. A. Yaroshevsky, and N. P. Ilyin (1971) A physicochemical model for element separation in the differentiation of mantle material, *Philos. Trans. R. Soc. London, Ser. A, 268,* 409–421.

Vogt, J. H. L. (1904) Die Silikatschmelzlösungen, II, Über die Schmelzpunkt-Erniedrigung der Silikatschmelzlösungen, *Videnskabs-Selskabets Skrifter, I, Math.-Naturv. Klasse (Christiania),* No. 1, 235 p.

von Cotta, B. (1858) *Geologische Fragen,* J. A. Engelhardt, Freiberg, 344 p.

Wachtman, J. B., Jr. (1974) Highlights of progress in the science of fracture of ceramics and glass, *J. Am. Ceram. Soc., 57,* 509–519.

Wager, L. R. (1958) Beneath the earth's crust, *Adv. Sci., 15,* 31–45.

Wager, L. R., G. M. Brown, and W. J. Wadsworth (1960) Types of igneous cumulates, *J. Petrol., 1,* 73–85.

Wagner, P. A. (1928) The evidence of the kimberlite pipes on the constitution of the outer part of the earth, *S. Afr. J. Sci., 25,* 127–148.

Wakita, H., H. Nagasawa, S. Uyeda, and H. Kuno (1967) Uranium, thorium and potassium contents of possible mantle materials, *Geochem. J., 1,* 183–198.

Waldbaum, D. R. (1971) Temperature changes associated with adiabatic decompression in geological processes, *Nature (London), 232,* 545–547.

Walsh, J. B. (1969) New analysis of attenuation of partially melted rock, *J. Geophys. Res., 74,* 4333–4337.

Warner, R. D. (1973) Liquidus relations in the system $CaO-MgO-SiO_2-H_2O$ at 10 kb P_{H_2O} and their petrologic significance, *Am. J. Sci., 273,* 925–946.

Washington, H. S. (1917) Chemical analyses of igneous rocks, *U.S. Geol. Surv. Prof. Pap. 99,* 1201 p.

Washington, H. S. (1925) The chemical composition of the earth, *Am. J. Sci., 9,* 351–378.

Weertman, J. (1968) Bubble coalescence in ice as a tool for the study of its deformation history, *J. Glaciol., 7,* 155–159.

Weertman, J. (1972) Coalescence of magma pockets into large pools in the upper mantle, *Geol. Soc. Am. Bull., 83,* 3531–3532.

White, W. P. (1909) Melting point determination, *Am. J. Sci., 28,* 453–489.

Whitehead, J. A., Jr., and D. S. Luther (1975) Dynamics of laboratory diapir and plume models, *J. Geophys. Res., 80,* 705–717.

Whitten, C. A. (1956) Crustal movement in California and Nevada, *Trans. Am. Geophys. Union, 37,* 393–398.

Wickman, F. E. (1966) Repose period patterns of volcanoes: I, Volcanic eruptions regarded as random phenomena; II, Eruption histories of some East Indian volcanoes; III, Eruption histories of some Japanese volcanoes; IV, Eruption histories of some selected volcanoes; V, General discussion and a tentative stochastic model, *Ark. Mineral. Geol., 4,* 291–367.

Wiederhorn, S. M., and L. H. Bolz (1970) Stress corrosion and static fatigue of glass, *J. Am. Ceram. Soc., 53,* 543–548.

Wilshire, H. G., and J. W. Shervais (1975) Al–augite and Cr–diopside ultramafic xenoliths in basaltic rocks from western United States, *Phys. Chem. Earth, 9,* 257–272.

Wilson, J. T. (1963) Continental drift, *Sci. Am., 208,* 86–100.

Wood, B. J., and S. Banno (1973) Garnet–orthopyroxene and orthopyroxene–clinopyroxene relationships in simple and complex systems, *Contrib. Mineral. Petrol., 42,* 109–124.

Wood, J. A. (1963) On the origin of chondrules and chondrites, *Icarus, 2,* 152–180.

Wright, T. L., W. T. Kinoshita, and D. L. Peck (1968) March 1965 eruption of

Kilauea volcano and the formation of Makaopuhi lava lake, *J. Geophys. Res., 73,* 3181–3205.

Wyllie, P. J. (1970) Ultramafic rocks and the upper mantle, *Mineral. Soc. Am. Spec. Pap. 3,* 3–32.

Yang, H.-Y., J. F. Salmon, and W. R. Foster (1972) Phase equilibria of the join akermanite–anorthite–forsterite in the system $CaO-MgO-Al_2O_3-SiO_2$ at atmospheric pressure, *Am. J. Sci., 272,* 161–188.

Yoder, H. S., Jr. (1952) Change of melting point of diopside with pressure, *J. Geol., 60,* 364–374.

Yoder, H. S., Jr. (1958) Effect of water on the melting of silicates, *Carnegie Inst. Washington Yearb. 57,* 189–191.

Yoder, H. S., Jr. (1965) Diopside–anorthite–water at five and ten kilobars and its bearing on explosive volcanism, *Carnegie Inst. Washington Yearb. 64,* 82–89.

Yoder, H. S., Jr. (1968) Experimental studies bearing on the origin of anorthosite, *N.Y. State Mus. Sci. Serv. Mem. 18,* 13–22.

Yoder, H. S., Jr. (1969) Calcalkalic andesites: experimental data bearing on the origin of their assumed characteristics. In A. R. McBirney, ed., Proceedings of the Andesite Conference, *Oreg. Dep. Geol. Miner. Ind. Bull. 65,* 77–89.

Yoder, H. S., Jr. (1973) Melilite stability and paragenesis, *Fortschr. Mineral., 50,* 140–173.

Yoder, H. S., Jr. (1974) Garnet peridotite as a parental material for basaltic liquids, *Carnegie Inst. Washington Yearb. 73,* 263–266.

Yoder, H. S., Jr. (1975a) Heat of melting of simple systems related to basalts and eclogites, *Carnegie Inst. Washington Yearb. 74,* 515–519.

Yoder, H. S., Jr. (1975b) Relationship of melilite-bearing rocks to kimberlite: a preliminary report on the system akermanite–CO_2, *Phys. Chem. Earth, 9,* 883–894.

Yoder, H. S., Jr., and I. Kushiro (1969) Melting of a hydrous phase: phlogopite, *Am. J. Sci., Schairer Vol., 267A,* 558–582.

Yoder, H. S., Jr., D. B. Stewart, and J. R. Smith (1957) Ternary feldspars, *Carnegie Inst. Washington Yearb. 56,* 206–214.

Yoder, H. S., Jr., and C. E. Tilley (1957) Basalt magmas, *Carnegie Inst. Washington Yearb. 56,* 156–161.

Yoder, H. S., Jr., and C. E. Tilley (1961) Simple basalt systems, *Carnegie Inst. Washington Yearb. 60,* 106–113.

Yoder, H. S., Jr., and C. E. Tilley (1962) Origin of basalt magmas: an experimental study of natural and synthetic rock systems, *J. Petrol., 3,* 342–532.

Yokoyama, I. (1957) Energetics in active volcanoes, 2nd paper, *Bull. Earthquake Res. Inst. Tokyo Univ., 35,* 75–97.

Yoshiki, B., and R. Yoshida (1952) Composition of low-alkali glass, *J. Am. Ceram. Soc., 35,* 166–169.

Yund, R. A. (1974) Coherent exsolution in the alkali feldspars. In A. W. Hofmann, B. J. Giletti, H. S. Yoder, Jr., and R. A. Yund, eds., Geochemical transport and kinetics, *Carnegie Inst. Washington Publ. 634,* 173–183.

Zemansky, M. W. (1937) *Heat and thermodynamics,* McGraw-Hill Book Company, New York, 388 p.

Author Index

237

Subject Index

The student of petrology may find it useful to formulate his own views on the concepts and principles summarized by a word or phrase and define the terms in The Subject Index prior to scholastic examination.

243

Magma norm
  variation with degree of melting, 152, 153
  volatile influence, 153
Magma production
  frequency, 191, 192
  rate of, 199
Magma rise
  forces causing, 186–189, 207, 208
  liquid content change, 203
Magma reservoirs
  depth of, 50
Magma separation
  factors involved, 133
  influence on magma type, 120
Magma source
  duration, 192
  model of, balloon/soda-straw, 3
  model of, Benioff-zone, 5
  model of, crystal-mush, 4
  model of, depth-of-generation, 7
  model of, explosive-diatreme, 9
  model of, geophysical, 4
  single, 118
  under volcano, 9
Magma Tap Program, iv
Magma type
  depth of generation, 9, 132
  determining factors, 204, 205
  eutecticlike behavior, 115
  from different layers, 7
  number of invariant points, 154
  principal control, 120
  related to invariant points, 205
  role of volatiles, 85, 205
Magma withdrawal
  beginning of melting, 134
  related to yield stress, 191
Magma zone
  life of, 118
Major phases
  restriction in kinds of, 202
Makaopuhi, Hawaii, 112, 190
Mantle
  amount molten, 51, 52
  composition, 42
  content of diopside, 84
  crystalline, seismic evidence, 12
  crystallinity, 49, 50
  degassing of, 84
  density, 5, 41

  depletion, depth of, 48
  differentiation, 101, 201
  gas composition, 84
  gas phase presence, 81
  heterogeneity, 10, 42, 111, 112, 201
  homogeneous, 8
  hydrous melting of, 81, 86
  metamorphosed, 201
  metasomatism, 84
  oceanic *vs.* continental, 51
  plastic zone, 180
  "plum pudding" concept, 43
  remelting of, 42
  seismic velocity, 5, 52
  shear in, 168
  state of stress, 61
  suboceanic, 41
  tensile strength, 171
  turnover, 48
  volatile content, 203
  volatile depletion, 84, 86
  water content, 84
  water flux, 84
Mantle material
  achondrite, 24
  carbonaceous chondrite, 25
  characteristic property, 115
  chondrite, 24
Mauna Loa, Hawaii
  volume of volcano, 193
Mechanical energy
  converted to heat, 71, 203
Mechanical instability
  crystal structure, 87
Megacrysts, 119
Melilite assemblages, 122
Melilite nephelinite
  rare-earth element abundances, 152
  related to degree of melting, 151
Melt
  amount generated in nuclear explosion, 184
  distribution in rock, 164
  effect of stress on, 207
  films, 52
  localized, 7
  migration of, 165
  phase contribution to, 206
  primordial residuum, 48
  seismic velocities, 52
  source region, 61
  structural state, 86, 87

System Index

References only to oxide and mineral end-member systems are included. Experimental studies on natural or simulated rocks are listed in the Subject Index. Mineral end-member systems related to the expanded basalt tetrahedron are listed in Tables 7-1 (p. 123), 7-2 (p. 125), and 7-3 (p. 125). Abbreviations are alphabetical and as used in text.

MINERAL END-MEMBER SYSTEM

Ab–H$_2$O, 79, 80
Ab–An–Di–Fo, 14
Ab$_{50}$An$_{50}$–Fo–Qz–H$_2$O (with 10% KAISi$_3$O$_8$), 38
Ab–Di–En–Qz, 123, 126–128
Ak–An–Di–Fo–Ne, 131
An–Di, 91, 92
An–Di–Fo, 16, 134
An–Di–H$_2$O, 199
An–Fo, 25
CaTs–En, 143
CaTs–Fo–SiO$_2$, 147
Cor–Di–En, 143, 144
Cpx$_{50}$Gr$_{50}$–Fo, 117
Di–CO$_2$, 78, 81
Di–CO$_2$H$_2$O, 83
Di–Fo–Qz, 36
Di–Fo–Qz–H$_2$O, 36, 37
Di–Fo–Py, 96, 107, 134, 170
Di–H$_2$O, 78, 198
Di–Py, 94, 149

En–H$_2$O, 198
Fa–Fo, 14, 93
Fo–H$_2$, 46, 47
Fo–H$_2$O, 198
Fo–Ks–Qz–H$_2$O, 38, 39
Fo–Ne–Qz, 132, 133, 147
Fo–Ne–Qz–CO$_2$, 133
Fo–Ne–Qz–H$_2$O, 133
Ph–H$_2$O, 82
Py–H$_2$O, 198

OXIDE SYSTEMS

Na$_2$O–CaO–Al$_2$O$_3$–SiO$_2$, 131
Na$_2$O–CaO–MgO–Al$_2$O$_3$–SiO$_2$, 121, 122
K$_2$O–SiO$_2$–H$_2$O, 196
K$_2$O–MgO–Al$_2$O$_3$–SiO$_2$, 131
K$_2$O–MgO–Al$_2$O$_3$–SiO$_2$–H$_2$O, 82
CaO–MgO–SiO$_2$, 165
CaO–MgO–Al$_2$O$_3$–SiO$_2$, 31, 33, 131, 145, 147
CaO–FeO–Al$_2$O$_3$–SiO$_2$, 131

265